Navigating Challenges for Sustainable Growth

Suman Bery · Kapil Kapoor · Jean-Louis Arcand
Editors

Navigating Challenges for Sustainable Growth

Insights from the Indian G20 Presidency

Springer

Editors
Suman Bery
NITI Aayog
New Delhi, Delhi, India

Kapil Kapoor
IDRC-Asia
New Delhi, Delhi, India

Jean-Louis Arcand
Global Development Network
New Delhi, Delhi, India

ISBN 978-981-97-7896-6 ISBN 978-981-97-7894-2 (eBook)
https://doi.org/10.1007/978-981-97-7894-2

This Springer imprint is published by the registered company Springer Nature Singapore Pte Ltd.
The registered company address is: 152 Beach Road, #21-01/04 Gateway East, Singapore 189721, Singapore

Foreword

A Green and Sustainable Growth Agenda for the Global Economy

The genesis of the G20 was at the end of the last century. Its salience grew even more during the global economic crisis in the first decade of the twenty-first century. Although India itself has had the scale and autonomy to bounce back relatively quickly from COVID, this is not true of most countries of the Global South where the development momentum has appreciably deteriorated because of the pandemic, further complicated by the aftershocks of war, sanctions, and inflation. The challenges of adaptation to a changing climate add to economic and social stress. There is a deep crisis of economic development, well-documented in the stagnation of the Sustainable Development Goals of the UN and in commentary by the International Monetary Fund and the World Bank, but for reasons of a changed geopolitics, this deep crisis has not evoked the solidarity of the past.

The G20 Leaders' most basic commitment is to deliver strong, sustainable, balanced, and inclusive growth, to be achieved through political guidance arrived at through dialogue and policy coordination. Each G20 Presidency builds upon the achievements of its predecessor and hands over to its successor through a well-established troika process. With Indonesia leading India, Brazil scheduled to follow in 2024 and South Africa in 2025, four developing countries in succession will hold the G20 Presidency. This sequence provides a golden opportunity to bring sustained long-term economic growth (and not just economic recovery) back to center stage on the global economic agenda.

On 28–29 July 2023, NITI Aayog, together with the International Development Research Centre (IDRC), Ottawa and the Global Development Network (GDN), New Delhi, convened an international policy conference of around 40 leading thinkers to examine prospects and challenges pertaining to green and sustainable growth for the global economy. The policy conference was an official G20 side event designed to explore the contours of a new growth model to guide this and future G20 Presidencies. The first day of the conference focused on themes related to energy, climate, and growth; technology, policy, and jobs; the growth implications of a fractured trading system and reshaping global finance for sustainable growth. The second day addressed themes related to multilateralism as well as adjustment, resilience, and inclusion in an uncertain world.

The discussions yielded rich insights in a number of areas. For instance, opinions are sharply divided on the growth consequences of a transition from fossil to renewable resources at a time when climate change will impose its own challenges of adaptation. The International Monetary Fund and the World Trade Organization have warned of the economic welfare consequences of the balkanization of international trade in the pursuit of economic security, at a time when interdependence through both trade and finance has become weaponized. There is increasing concern that, for all its sophistication, the global finance system, both official and private, is not a force supporting long-term sustainable growth. In various sessions of the conference, experts addressed the critical issue of how the global financial order (including the monetary order) should be reformed to be more supportive of rising living standards across the world. Further, they deliberated upon whether a division of world trade into regional blocs is inevitable and what this implies for a great majority of the world's countries.

The issue of liberal globalization and rising inequality at least within nations, was also discussed. As the G20 countries lead a global debate on the development strategy, ensuring more inclusive outcomes will be of paramount concern for both developing and developed countries. Many experts highlighted the critical role played by treaty-based multilateral institutions designed at the end of the Second World War for securing important global public goods (peace and economic growth) and the pathways for reforming multilateralism in the present era.

The discussions held during the policy conference offered several important suggestions for green and sustainable growth which NITI is pursuing through various fora. By design the conference preceded the Leaders' Summit, allowing some of the key ideas and thoughts shared by the experts during the 2-day deliberations to be introduced in the unanimous Leaders' Declaration, a great achievement of the Indian Presidency under Prime Minister Modi's leadership.

The insights shared by leading experts during the conference can serve as invaluable benchmarks for researchers, academics, and stakeholders globally. The conference sessions received an overwhelming response from across the world, indicating tremendous interest among the public and academia in the subject matter. This brochure is our collective contribution to the body of knowledge on this subject, and we hope that it will provide valuable inputs for Brazil as it takes over the G20 Presidency from India.

Under Prime Minister Modi's leadership and with his strong personal engagement, India took its G20 Presidency very seriously. The *Sabka Saath Sabka Vikas* model that has shown the way in India can also be a guiding principle for the welfare of the world.

In producing a consensus document at a time of great division between major powers, India succeeded in imparting new purpose and momentum to a grouping that had been in danger of irrelevance. At its core, the New Delhi Leaders' Declaration is cross-cutting and driving transformative action in an integrated manner. A collective vision on economic and social empowerment, bridging the digital divide, driving gender-inclusive climate action, and securing women's food security, nutrition, and well-being make the Declaration the most ambitious communique in terms of driving gender equality and women-led development. The Global South's call for enhanced representation and voice in multilateralism as a whole was brought forward, especially through the inclusion of the African Union into the G20 as a permanent member.

India's G20 Presidency has focused extensively on working towards better, bigger, and more effective Multilateral Development Banks (MDBs) that can deliver finance and support for twenty-first-century issues. India has taken on the MDB initiative from a start made by Indonesia on the capital adequacy framework. As Brazil takes over from India, given the sophistication and integration of its own capital markets, it will take on the task of reshaping the skills and talents of private finance to serve the development challenges of a hurting planet.

(Suman Bery)
Vice Chairman
NITI Aayog

(Amitabh Kant)
G20 Sherpa
G20 Secretariat

Preface

A Green and Sustainable Growth Agenda for the Global Economy

The G20 has evolved to unite member states beyond its original goal of international economic cooperation, but to address the world's most complex and pressing challenges. The consecutive presidencies of emerging economies including Indonesia, India, Brazil, and South Africa, demonstrate a commitment to address these evolving challenges and advance a development agenda shaped and led by the Global South.

India's G20 Presidency has created several opportunities to spearhead southern solutions, which has led to the emergence of a more inclusive and equitable framework for the G20. Building on this momentum created by India, NITI Aayog with support from the International Development Research Centre (IDRC), and in collaboration with the Global Development Network (GDN), set the stage for a global conversation through a conference on A Green and Sustainable Growth Agenda for the Global Economy, which was held in New Delhi in July 2023.

The event embodied the voices, vision, ambition, and collaborative spirit of the Global South to find solutions to address the multiple crises and emerging challenges faced by the world. The recommendations that arose from this conference will also play a critical role in shaping the dialogue and will help build consensus among member states under the subsequent presidencies of Brazil and South Africa, ensuring continuity and relevance of the discourse.

This conference was a confluence of ideas and perspectives, spread across six sessions and a keynote address, aiming to unravel complex global issues and forge pathways for an inclusive, just, and sustainable future. The emphasis on a 'just transition' as a critical pathway to mitigate climate change marked a pivotal point in the discussions, underlining the potential positive economic impact for developed and emerging economies, such as India, in their journey towards net-zero emissions. This transition, albeit fraught with challenges such as financial constraints and technological needs, calls for a reframing of the global financial architecture and norms.

We are navigating through times marked by technological disruptions, geopolitical realignments, and mounting environmental pressures. The conference underscored the transformative power of technology, particularly Digital Public Infrastructure, in reshaping economies and the job market. It highlighted the shift from hyper-globalisation to slow-balisation, pointing towards the need for an equitable transition in the global economy and financial systems.

Furthermore, the discussions delved deep into the evolving nature of capital, the changing dynamics of labour markets, and the complex interplay between economics, domestic politics, and geopolitics. The conference also brought to fore the urgent need to address disparities in economic recovery and the crucial role of effective multilateralism.

The expert-led discussions also offered a realistic picture of the international development landscape, acknowledging uneven progress and highlighting the challenges and opportunities.

This report aims to encapsulate the essence of the discussions, the urgency of the challenges, and the collective stride towards solutions. It offers an incisive analysis of the discussions at each session and aims to foster a comprehensive understanding of the conference's outcomes and its implications for future G20 presidencies, and the world at large.

Kapil Kapoor
Regional Director for Asia
IDRC
New Delhi, India

Jean-Louis Arcand
President
Global Development Network
New Delhi, India

Executive Summary

The conference was organised around six sessions and one keynote address. In what follows, we summarise the main points that emerged, both from the presentations and the discussions that followed.

Session 1. Energy, Climate, Growth

It has become patently obvious that there is a need for a just transition to mitigate climate change. But there is an upside: for example, there are potentially positive economic impacts of India's transition to net-zero emissions, which involves a shift away from fossil fuel imports, which could improve the country's balance of payments. Constraints include financial resources and technology, necessitating a reconstruction of the global financial architecture.

Challenges. Speakers identified four overarching challenges: global cooperation in climate economics, India's specific challenges and opportunities in transition, trade-offs in the political economy of climate action and viewing climate adaptation as an economic problem. Climate change benefits are distributed globally, but costs are localised, necessitating international collaboration. This is particularly crucial considering the upfront costs of mitigation compared to the long-term impacts. Strategies such as carbon pricing, smart infrastructure, and the flexible elements of the Paris Agreement can help manage these costs. As a significant global emitter reliant on fossil fuels, India faces a challenging transition. However, renewable energy sources are gaining ground. Notably, the transition presents substantial potential for job creation, particularly through distributed energy and new economic activities.

Climate emergencies exacerbate fiscal challenges for countries such as India. Other significant trade-offs include energy access versus clean energy; energy security versus energy sustainability; and job growth versus job losses in the transition. Finally, and given the high climate variability and vulnerability of regions like

India, there is a need to view adaptation as an economic issue. This includes incentivising transitions in individual choice, procurement, infrastructure programmes, and innovation.

Way Forward. A two-pronged approach was proposed going forward. First, multilateralism is a necessity (and not an option). Second, flexibility is central to the Global South. Climate change's challenges require multilateral solutions, including creating resilience funds, de-risking platforms, promoting circular economies, joint technological development, and enhancing green energy security and transition partnerships. However, to cope with the disproportionate impacts of climate change, setting up a 'flexibility mission' is important for these countries. This would provide them with the means to adapt and innovate in response to environmental shifts.

Session 2. Technology, Policy, Jobs

The current state of the world is characterised by technological disruptions, global realignments, and environmental pressures. First, the emergence of new economic models and intricate national security concerns are linked to the ongoing technological revolution, particularly advances in AI. Second, demographic and economic shifts are contributing to the rise of a multipolar world. Third, the strain from global growth on the environment is leading to significant impacts on natural systems, raising the likelihood of unforeseen, disruptive events.

Challenges. The challenges are four-pronged. First, the world faces major job market transformations: technological progress and AI are creating a structural labour market churn, reshaping the job landscape, and raising concerns about job displacement. Second, we face significant demographic concerns: the ageing global population presents a significant challenge to maintaining economic growth. Third, we will face increased environmentally driven migration: migration triggered by environmental pressures could shift future carbon emissions patterns. Finally, numerous policy dilemmas will be caused by advances in gen-AI: the rapid pace of AI progression is creating numerous policy challenges including regulatory issues, AI's carbon footprint, ethical considerations, security risks, and potential job losses.

Way Forward. The way forward involves investing in human capital, ensuring that sustainability is focused on wellbeing, reimagining multilateral cooperation, regulating and constructing governance structures for AI, building inclusive technological infrastructure and embracing new labour practices and emerging technologies. First, it is crucial to invest in education and stimulate labour force participation through measures such as extending the retirement age. Second, sustainability should be viewed broadly as encompassing human wellbeing rather than just economic activity. Third, there is a need to strengthen and restructure international cooperation to tackle shared challenges such as skill development, job creation, and productivity. Fourth, generative AI requires a comprehensive regulatory framework, ethical guidelines, and proactive industry self-governance. Fifth, investing in technological infrastructure, enhancing digital education, and boosting public-private partnerships are essential for an inclusive and sustainable transition. Finally, adopting progressive labour practices and leveraging emerging technologies, particularly in developing markets, can be transformative. Industry investment in R&D, training, and multidisciplinary research collaborations are also key to navigating future challenges.

Keynote Address by Nandan Nilekani

In an era of global demographic shifts, warming climates, and the dawn of new geopolitical contexts, technology is enabling the restructuring of economies. In India, this evolution has involved a transition from a predominantly offline, informal, and low-productivity landscape to a unified, formal mega-economy underpinned by Digital Public Infrastructure (DPI). DPIs, funded either publicly or driven privately through regulatory policies, enable interoperability and create combinatorial benefits.

Challenges. India's diversity, in terms of cultures, markets, industrialisation levels, and regulations posed significant challenges to economic formalisation. Informality and the lack of productive engagement with technological advancements rendered the economy vulnerable to inefficiencies. The need for rapid financial inclusion, transactional formalisation, and a dynamic startup ecosystem was acute. DPIs needed to balance the tension between fostering innovation and ensuring robust regulation. The impact of climate change further necessitated the need for mechanisms that could expedite both mitigation and adaptation efforts.

Way Forward. DPIs have been pivotal in transforming India's economic landscape. They have accelerated financial inclusion, facilitated the highest volume of digital payments globally, and provided the foundation for the economy's formalisation. By embedding policy into source codes, DPIs reconcile the demand for innovation with regulatory necessities. Welfare schemes and programmes such as distress-related money transfers and vaccination initiatives have become more efficient due to these systems. The startup ecosystem has also flourished, surging from around a thousand startups in 2016 to 115,000 currently. DPIs such as the Open Network for Digital Commerce (ONDC) have democratised the digital economy, fostering competitive and equitable market dynamics. Furthermore, DPIs can play a crucial role in climate change action, aiding in anticipatory climate financing, the promotion of a circular economy, and the creation of energy interfaces. Through such multi-pronged interventions, DPIs are shaping the contours of a more resilient, equitable, and sustainable economy.

Session 3. Growth Implications of a Fractured Trading System

Technological innovations and transportation advancements led to the fragmentation of manufacturing processes, reducing trade barriers and incorporating a billion lower-wage workers into the global labour supply. This era was marked by a rise in global GDP and trade. This produced the age of hyper-globalisation. But *a* noticeable shift has occurred towards 'slow-balisation' or de-globalisation, characterised by increased restrictions on trade, labour movements, and limited technology diffusion. This transition indicates a significant transformation in the global economy.

Challenges. Two main challenges were highlighted: China's resilience and dominance and the impact of de-globalisation on Europe and US-China relations. Despite the global slowdown, China has maintained its position as a key driver of global growth. China's resilience is largely attributed to its focus on exporting more value-added products, particularly in clean tech. Conversely, Europe has suffered significant losses due to de-globalisation. Concurrently, the current US-China decoupling has led to increased trade and financial protectionism. The world's increasing dependence on China, particularly in clean and green tech sectors, suggests potential risks for global supply chains.

Way Forward. The way forward involves India as a potential future engine of growth, the adoption of innovative strategies centred on sustainable growth and the need for trade reform. With China facing an ageing population, India could step up as the next global growth engine. However, it must address its de-industrialisation and boost its manufacturing sector. Strategies should include enhancing productivity in traditional sectors, creating manufacturing jobs, managing competitive exchange rates, closing infrastructure and logistics gaps, and improving education and skills development. Initiatives such as the Gati-Shakti Masterplan and a potential Universal Basic Income (UBI) scheme could help sustain India's growth momentum. To facilitate this transition and maintain growth momentum, reforms of multilateral trade rules are needed. These would retrieve policy space for industrialisation and facilitate the transfer of technology.

Session 4. Reshaping Global Finance for Sustainable Growth

The current global financial architecture, marred by dysfunction and cluttered with non-economic issues, requires significant restructuring. It is instrumental in financing growth, which currently faces a shortfall of $3 trillion over the next decade, highlighting an urgent need for both public and private resource mobilisation. Additionally, the fragmented nature of this architecture is contributing to disparities in economic recovery between developed and developing regions. These challenges call for an orderly transition to more efficient and sustainable systems.

Challenges. Increasing global financing needs in the face of dwindling liquidity present a substantial challenge, particularly for developing nations. Three types of financing—private sector, multilateral, and bilateral—each come with their own complications, from volatility to limited availability. Emerging economies are further threatened by capital volatility and exchange rate risks as their financial integration deepens. The global debt architecture, currently informal, inefficient, and fragmented, poses additional challenges, with many low-income countries already in or nearing a debt crisis. Moreover, the world lags in transitioning to net-zero emissions, exacerbating the climate crisis and necessitating significant, upfront financing, primarily from private sources.

Way Forward. Reforming financial systems and processes is key to overcoming these challenges. This includes making SDR allocation rule-based and less discretionary, as well as improving the multilateral system. Debt needs to be managed sustainably and transparently, possibly by establishing a multilateral creditor club and strengthening legal frameworks. Financial safety nets need strengthening, bilateral swap lines need to be expanded, and IMF contingency lines invoked to make capital flows safer. The role of credit rating agencies should also be regulated to ensure fair assessments for emerging countries. Finally, scaling up green financing, especially for regions like Africa, will require innovative policies and tools to drive the global transition to net-zero emissions. This includes fostering green investors, policies that bolster the enabling environment, and encouraging international cooperation.

Virtual Discussion and Wrap-Up of Day 1

The nature of capital has evolved considerably since the 1980s, shifting from tangible to intangible forms such as software, databases, and patents. Alongside this, a skill-biassed form of technological change has led to diminished manufacturing employment and a polarised job market. Future opportunities lie in the globalisation of knowledge, fintech, e-commerce, trade in services, and remote work. Green growth strategies are also emerging, although they will create both winners and losers in the labour market.

Challenges. The main challenges involve technological disparities and the inefficiency of the MDBs. The rise of skill-intensive technologies and labour-saving innovations, such as robotics and AI, pose significant challenges for less skilled workers and countries specialising in labour-intensive processes. Simultaneously, the Multilateral Development Banks (MDBs), though critical in supporting socio-economic development, are struggling with significant performance gaps, transparency issues, and the impacts of geopolitics. Their effectiveness and efficiency are being undermined, leaving a sizable void in the financial and technical support required by most developing countries.

Way Forward. The way forward involves a new toolkit of policies and a reform of the structure of the MDBs. Addressing these challenges necessitates new policies promoting advanced manufacturing and skill-intensive technologies such as semiconductors. Efforts should be directed towards ensuring macroeconomic stability and promoting inclusive growth, alongside effective transfer, adjustment, and training policies. Concomitantly, MDBs must engage in transformative reforms, including improved capital mobilisation, better project implementation, joint financing, risk sharing, and making sustainable infrastructure an asset class. As part of a new multilateralism, MDBs must not only improve their own functionality but also ramp up public and private investments in developing countries, which are essential to meeting global challenges such as climate change.

Session 5. Multilateralism: Geopolitics, Governance, and the Global Commons

The role of economics in politics and the effects of geopolitics on multilateralism underscores the interdependence of global issues. These require an effective multilateral order for universal benefits, encompassing growth, sustainable and inclusive development, peace, and risk management. Important factors include open trade and investment, the management of global tensions, such as the US-China trade war, and the integration of emerging economies like India into the world economy. Understanding health as both a consumption good and an investment allows a focus on universal health coverage, malaria eradication, and new vaccines. However, the relationship between health and economics is complex, particularly given the disruption caused by COVID-19.

Challenges. The rise of protectionism and violations of non-discrimination rules have strained multilateralism, triggering a shift from a unipolar to a multipolar world. Systemic crises stemming from disruptive technological innovations and the effects of the COVID-19 pandemic present significant hurdles. Severe recessions transmitted from the Global North to the Global South through trade channels highlight the harmful economic spillover. This spillover, coupled with a decrease in official development assistance, has prompted calls for an effective insurance mechanism. Furthermore, the multitude of Sustainable Development Goals (SDGs) can be overwhelming; a focused approach may yield better results. Misinformation regarding COVID-19 has also highlighted the importance of examining the role of institutions.

Way Forward. In addressing these challenges, key recommendations include reforming the World Trade Organization, promoting plurilateral initiatives, regional trade agreements, and comprehensive partnerships. Moreover, leveraging the G20 to represent the Global South and advocating for more manageable, tangible goals within the SDG framework could provide beneficial outcomes. Simultaneously, there is a need to overhaul multilateral institutions and encourage the growth of 'minilaterals' such as regional organisations, global NGOs, and big tech companies. As we grapple with reducing dependence on the global commons for developing nations' growth aspirations, it is crucial to bridge natural and social sciences and consider flexible rules within the global order. This will necessitate multinational corporations to meet certain thresholds to be included in multilateral arrangements. The need for a multilateral arrangement supporting global health goals and even the concept of an insurance mechanism for the Global South against shocks to the Global North given their interdependence is also evident, emphasising a balanced, adaptable approach to the evolving global order. There is also a scope for a trust fund type mechanism with leveraging so that the use of public funds meaningfully involves donors, recipients, and all stakeholders. Future multilateral partnerships should not have permanent membership with veto power, and should involve other stakeholders whose voices are sometimes drowned out such as the corporate sector, NGOs, and faith-based organisations.

Session 6. Adjustment, Resilience, and Inclusion in an Uncertain World

The international development landscape presents a complex picture, with uneven progress across different regions and demographic groups. Sub-Saharan Africa (SSA), despite showing some decrease in poverty rates, is trailing behind the global average in poverty reduction, owing to its low elasticity of poverty reduction with respect to economic growth. By 2100, almost half of the world's youth will be in Africa, creating both an opportunity and a jobs challenge. Concomitantly, robust economic growth in developing countries such as China and India has started to lower global inequality, though low-income countries are still lagging.

Challenges. Various challenges impede efforts towards full inclusion and equitable growth. In SSA, most workers moving out of agriculture end up in the informal sector, indicating weak patterns of structural transformation. Despite a decline in poverty, Latin America grapples with high inequality and deep poverty due to a dual social insurance architecture that does not adjust well to labour market dynamics. This model creates a trade-off between enhancing benefits for informal sector workers and maintaining productivity. Similarly, although improvements are observed in India's multidimensional poverty index, an annual increase in poverty numbers necessitates the development of an inclusion norm based on health, nutrition, and overall well being.

Way Forward. Building resilience and inclusion for subsector-level growth through product space mappings and value chain upgrades could help drive employment in Africa. Growing middle-income countries need to design their social protection systems to minimise problems and trade-offs associated with coverage expansion. Careful consideration of budgetary costs, redistribution, and efficiency is crucial when implementing social protection policies. A focus on education can boost economic inclusion, as demonstrated by unskilled workers' wage increases in countries like India. Finally, addressing multidimensional poverty by considering aspects beyond income, such as health and nutrition, could contribute to comprehensive and sustainable development.

Contents

About the Editors

Suman Bery is currently Vice Chairperson, NITI Aayog, in the rank and status of a Cabinet Minister. An experienced policy economist and research administrator, Mr. Bery took over as NITI Aayog Vice Chairperson from 1 May 2022. At the time of his appointment, Mr. Bery was a Global Fellow in the Asia Programme of the Woodrow Wilson International Centre for Scholars in Washington D.C. and a non-resident fellow at Bruegel, an economic policy research institution in Brussels. He was also a member of the Board of the Shakti Sustainable Energy Foundation, New Delhi. From early 2012 till mid-2016, Mr. Bery was Royal Dutch Shell's global Chief Economist based in The Hague. In this capacity, he advised the board and management on global economic and political developments. He was also part of the senior leadership of Shell's global scenarios group. During his time at Shell, he led a collaborative project with Indian think tanks (later published) to apply scenario modeling to India's energy sector.

Kapil Kapoor is the Regional Director for the Asia Regional Office of the IDRC. Kapil has over 30 years' experience in international development, specialising in Africa and Asia. He was the Director General for Southern Africa, at the African Development Bank, where he was responsible for the Bank's projects and programmes across 13 countries in Southern Africa. He has also served as the Director for Strategy and Operational Policies at the Bank, where he led the preparation of the Bank's Long-Term Strategy for Africa, and the Bank's Private Sector Development Strategy. Kapil has held a series of senior positions with the World Bank Group, including the World Bank's Representative for Uganda and Zambia and the World Bank's Sector Manager for its poverty reduction, economic management and governance programme in Asia. Kapil holds a Ph.D. degree in Economics and an MBA degree in Finance.

Prof. Jean-Louis Arcand a Canadian citizen born in Cameroon, is the President of GDN, Professor of economics at the Graduate Institute of International and Development Studies in Geneva, as well as an Affiliate Professor at the Université Mohammed VI Polytechnic in Rabat. He is a Founding Fellow of the European

Development Research Network (EUDN), a Senior Fellow at the Fondation pour les études et recherches en développement international (FERDI) and has been a Visiting Professor at Renmin University of China in Beijing, Universidade Federal da Bahia and several universities in Africa and the Caribbean. He was assistant and then Associate Professor at the University of Montréal, and Professor at the Centre d'études et de recherches en développement international (CERDI). Jean-Louis holds a Ph.D. in Economics from the Massachusetts Institute of Technology (MIT), an M.Phil. from Cambridge University and a B.A. (high honors) from Swarthmore College.

Chapter 1
Energy, Climate, Growth

Jayant Sinha, Robert Stavins, Jessica Seddon, and Arunabha Ghosh

1.1 Session Chair: Jayant Sinha

I think it makes sense to spend a few minutes framing the issues as we have seen them in India and globally. Thereafter, each of the panellists will shortly introduce themselves and their work, and spend a few minutes giving us their perspectives. They have presentations that they will be providing. Thereafter, we will have an opportunity to take questions from all of you. So, we look forward to a very lively session, and we will of course do a quick summing up. Each of the distinguished speakers will give their perspectives. I'll give my perspective, and we will conclude. The good news is we have plenty of time. It's a very rich and complicated set of issues, but we are very fortunate to have very distinguished commentators on that. And of course, all of you will provide your perspectives as well.

Now, before we get started, I do want to acknowledge our great enthusiasm and excitement about the fact that through this G20 process as the Sherpa described, there is an outcome that we are working towards, which is a Green Development pact. And the Pact would be a set of coordinated actions to really drive, as Sumanji just said, growth, particularly sustainable green growth. And when it comes to green

J. Sinha (✉)
Member of Parliament & Chair of Parliamentary Standing Committee on Finance, Delhi, India

R. Stavins
AJ Meyer Professor, Energy and Economic Development, Harvard University, Cambridge, MA, USA

J. Seddon
Senior Fellow, Yale Jackson School of Global Affairs, New Haven, USA

Artha Global, London, UK

A. Ghosh
CEO, Council on Energy, Environment and Water (CEEW), New Delhi, India

S. Bery et al. (eds.), *Navigating Challenges for Sustainable Growth*,
https://doi.org/10.1007/978-981-97-7894-2_1

and sustainable growth, I must say that I have a very personal and deep interest in that, as Sumanji was also outlining.

I represent Hazaribagh in Jharkhand and Hazaribagh is one of the largest coal producing areas in India and in fact in the world. So we are rich, we sit on top of one of the richest coal seams in the world, which is the north current coal field, which is where we are producing much of our coal now. And not only do we have a tremendous amount of coal production happening, we have in the area 6,000 megawatts of coal-fired capacity coming up in front of our eyes. We have a whole host of sponge iron plants. We have brick kilns, we have a large steel rolling mill also, in fact in my constituency. So, we have a full industrial cluster that is very fossil fuel dependent.

All of these issues of how do we ensure green development? How do we ensure just transition, how do we ensure that a big geographic area of five or 10 million people that is largely dependent on fossil fuel can actually undertake a just transition, are issues that I'm dealing with, not just at a high economic level, but day-to-day in terms of people's livelihoods, and all of the impacts of making that transition. So, these are issues of great interest to me, of course. I also, as Suman ji indicated, chair the parliamentary standing committee on finance, where we are of course looking at longer term issues of economic growth, financing of green development, of sustainable growth as well. What do we need to do both domestically and globally to ensure that this green development actually happens?

So, whether you look at it from a very grassroots perspective of actually the human impact of all of this, or we look at it from a macro perspective in terms of India's economic growth and the role of the global community in supporting that. Those are issues that I'm dealing with on a continuous basis now. I first started looking at these issues about four or five years ago in detail and looking at all the academic and modelling work that had been done in this area. As a policymaker, the first question I asked was, what's the impact of net zero? Because the conventional view at that time, particularly in India, was that net zero would actually be bad for India. If we embarked on a decarbonisation trajectory and we took India to net zero, it would actually be inimical. It would not be advantageous for India to actually do that. And I'm talking about this independently of the impact of climate change, purely on an economic and a business perspective. Is net zero positive or is net zero negative?

And what are India's emissions under different scenarios? Suman ji, of course, worked at Shell where they pioneered scenario analysis. So, from a scenario perspective, if we went to decarbonisation in different ways, how would that impact India's economic growth and India's economic outcomes? Therefore, I asked a number of economists.

I asked a number of think tanks. Show me the models, right? Help me understand what is the impact of decarbonisation. And what we found four or five years ago is that the climate scientists had done this for the US and China, but not for India. It was important to put together the climate models, the emissions models with what was happening as far as the economy was concerned. And by that I meant investments required for economic growth, jobs impact on balance of payments, and so on.

And by that I meant investments required economic growth, jobs impact on balance of payments and so on. So, we actually had to build a lot of the modelling work to be able to understand what's the impact of net zero. I think as all of you would intuitively accept and understand what we found, and not just through one modelling study, but a range of modelling studies done by a wide group of experts and think tanks, is that net zero?

And for a politician like me, the headline is important. The headline is very important for us as politicians. And for me the headline is net zero is net positive. Net zero is net positive. And I'm stripping out climate change. I'm not looking at the impact of extreme weather or 2.8-degree warming, any of that. I'm just saying purely on an economic and as a business question, is it better to move towards green technologies and decarbonisation or to stay with the fossil fuel driven economy? And the answer was independent of climate, purely on the basis of technology change and what was happening, solar getting cheaper than, you know, fossil fuels and so on.

Electric vehicles being from a Total Cost of Ownership (TCO) perspective, being cheaper than ice cars and so on, it's actually better to go to net zero. Net zero is net positive. There's no question about it. Once we factored in climate change, of course, then it becomes even more positive. But net zero is absolutely net positive when we look at it on every dimension, whether we look at it in terms of GDP growth, job creation, air pollution, balance of payments on all dimensions.

Net zero's, net positive. And let me just emphasise India's imports annually, $600 billion of which $250 billion are fossil fuels alone. So, if we can move away from fossil fuel imports, which is obviously crude oil, natural gas, coal, we are going to have a tremendously positive impact on our balance of payments as well. So therefore, net zero is net positive on every dimension that we look at. However, the challenge for everyone is how do we actually make it happen? Does the Global South have the financial resources and the technologies to actually move to net zero by 2060, 2070, which is when it might be realistic or practical to envision that it could happen? Do we have the money? And the answer is right now, neither do we have the financial resources nor the technology. And this requires a massive set of changes to the global financial architecture, to the MDBs and to the provision of global public goods as well. So those are the challenges that we face as we move towards net zero.

1.2 Speaker 1: Robert Stavins

My topic, as you can see, is a broad overview to get things started: "The Energy Transition: Challenges, Trade-offs, and Opportunities." I think of this as a primer for those of you who are not steeped in climate change policy, and are focused on economic development more broadly. So, I am trying to set a common denominator for discussion going forward. I am going to start in two ways: one spatial and one temporal, taking us from some basic science to basic economics to the geopolitics of climate change.

Starting with the spatial, greenhouse gases mix in the atmosphere, so the location has no effect on impacts in economic terms. Climate change is a global commons problem. What that means, economically, is that any jurisdiction that takes action will incur the costs of its actions, but the climate benefits will be distributed globally. If you think about the basic arithmetic of that, it becomes obvious that for virtually any jurisdiction, the climate benefits it reaps from its actions are going to be less than the costs it incurs, despite the fact that the global benefits of its actions might be much greater than the global costs of its actions. This presents a classic free rider problem, which is why international, but not global cooperation is essential. I say "international" partly because the G20 countries and regions alone account for, depending upon the accounting mechanism, 80 to 90% of global emissions.

Now, there is also a temporal dimension that takes us from the science to the economics to the politics and policy. Greenhouse gases accumulate in the atmosphere. Importantly, carbon dioxide (CO_2) has a half-life in the atmosphere of over a hundred years. This is in tremendous contrast with another very important greenhouse gas, methane, which we can talk about if there is interest. The damages of climate change are a function of the stock in the atmosphere, the concentration, not the flow at any point in time. The severe consequences of climate change are over this long time-horizon, but climate change policies and the attendant costs of mitigation are going to be upfront. For representative democracies, this presents a massive challenge of upfront costs and delayed benefits. The political incentive in democracies is to give benefits to voters today, and to place the costs on future generations. The climate problem is asking politicians to do precisely the opposite.

So, if we combine the global commons nature of the problem with this intertemporal asymmetry, that is essentially why, at least from an economic perspective, this is a very tough political challenge virtually anywhere in the world.

Having said that, I think the challenges are particularly striking for a country such as India. I say this, first of all, because India is a major force in the global energy economy. Energy consumption has more than doubled since the year 2000. This has been propelled both by a growing population and also by rapid economic growth. At the same time, however, energy use per capita in India is well under half of the global average.

Within the country, over 80% of energy needs are met by coal, oil, and solid biomass. Coal is by far the most important source of energy and has been critical for the expansion of electricity generation, the achievement of poverty reduction in the country, and for the expansion of industry. Oil consumption also has grown rapidly due to rising vehicle ownership and mass and private transport. And finally, biomass, primarily fuel wood, although it makes up a declining share of the energy mix, is still very widely used in India, particularly for cooking.

India is the third-largest global emitter of CO_2 as a result of all of this, despite the fact that it has relatively low per capita CO_2 emissions. Importantly, particulate emissions, particularly PM2.5, are a major health issue, correlated with CO_2 emissions but not causally linked with climate change. Finally, renewable sources of energy have begun to gain ground within the country. Solar photovoltaic (PV) growth has been nothing less than spectacular, and the potential for PV development and penetration going forward is even greater.

Trade-offs, therefore, going forward, are inevitable, and are best not ignored. It is best not to sweep them under the rug and pretend that it is all win–win "happy talk." The transition from fossil fuels to renewables will not be easy, and it will not be cheap. If it were either, it already would have happened. Costs of decarbonizing electricity, the grid, and transport sectors are going to be significant, but those costs can be reduced in various ways.

One way to keep costs down is through the judicious use of innovative policies. For example, there are carbon pricing instruments such as cap and trade, which India has already begun to do, improved pricing of electricity, including the use of smart metering where feasible, which has also begun to happen in this country. Costs can also be reduced through the judicious use of the flexible elements in international policy, that is, the Paris Agreement. In particular, appropriate use of the flexibility that is inherent in Article 6 of the Paris Agreement is, in my view, key. I think India could play a valuable leadership role in helping define the parameters of both Article 6.2 and Article 6.4.

It should also be recognized that reducing coal and oil use will, as I suggested a moment ago, bring tremendous health co-benefits. Typically, whether it is China, the United States, or India, those health co-benefits are actually of much greater value than are the local climate benefits. So they are very, very important scientifically, economically, and also potentially politically. Addressing climate change, importantly, is also going to reduce long-run costs of adaptation.

Now, let me just say very briefly why economists like myself tend to favor carbon pricing, taxes, or cap and trade in large, complex economies, but certainly not all countries of the world. It always depends upon national circumstances, but I am thinking of the G20 countries mainly. There are three reasons.

The first reason is feasibility: no other feasible approach can provide meaningful emission reductions. It is impossible to think about using conventional performance standards or technology standards when you have hundreds of millions of sources within the country.

Second, economically, it is the least costly approach in the short term because abatement costs are so heterogeneous.

And third, in the long term, carbon pricing can bring down costs because of providing incentives for carbon friendly technological change.

But I want to emphasize that although economists like myself may see it as necessary, eventually in large complex economies, it will not be sufficient. That is because there are other market failures; two in particular. One, there are some principal agent problems, and then there are also some public good issues, such as for information spillovers that get in the way of the pricing.

Now, in terms of the worldwide status of carbon pricing instruments, there are now cap and trade systems in place or announced in some very important countries of the world within the G20. There are also carbon taxes in many countries of the world, and we might want to compare them. There are equal numbers of the two, a total of about 60 recently, and approximately 30 of each. There are carbon taxes in particular Northern Europe that are at a much higher level than any of the cap and trade systems.

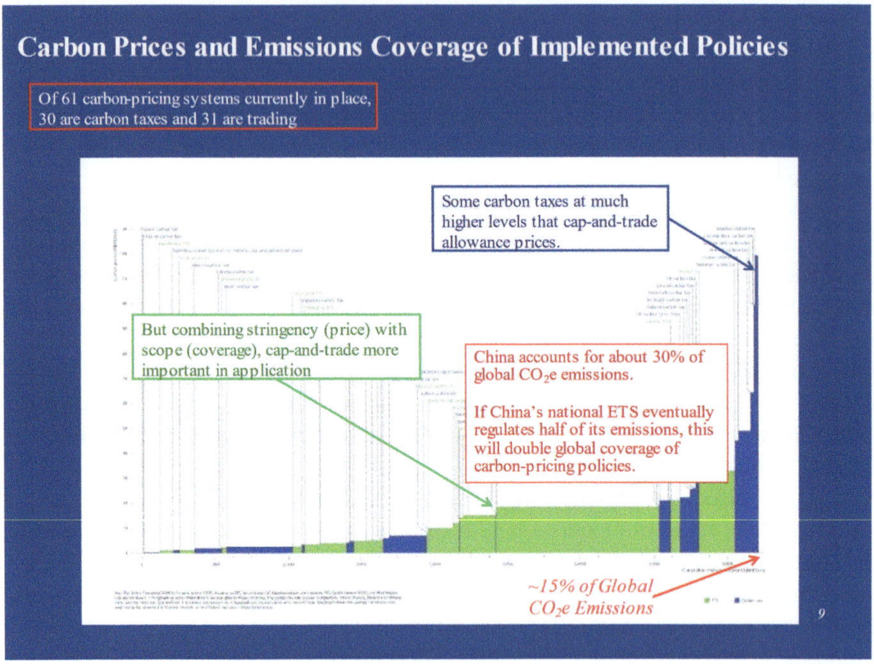

The taxes are in blue, the trading programs are in green. If we wanted to ask, which are more important right now, we could multiply the stringency, the carbon price, whether it is the tax or the allowance price from the market, by the scope, how many tons of emissions are accounted for, in other words, the area of the rectangles. And as you can see, there is more green than blue. So right now, emissions trading is more important in the world. I am not saying it will be in the future, compared to carbon taxes.

Altogether that is only 15% of global CO_2 equivalent emissions. Now interestingly, if China goes ahead, and they have now launched their emissions trading system, then eventually that would double the width of this because they account for 30% of global emissions. If they cover half of their emissions with their trading system, which eventually they say it will do, that would be 15% of global emissions.

Now, what are the consequences? First, the consequences for coal are very, very significant due to its high carbon content. This is not just for carbon pricing, but for any meaningful climate policy. Impacts are in terms of electricity dispatch, in terms of investments in new capacity, in terms of earlier retirement. In the case of natural gas, the impacts are smaller because of the lower carbon content, but in the short term, in countries where natural gas substitutes for coal in electricity generation, as it does in the United States, natural gas demand could increase in the short term. Having said that, it is important to recognize that even with substantial carbon pricing on the order of a hundred dollars per ton of CO_2, the effects are small compared to what exogenous technological change has meant, at least in the United States, because of the technologies of horizontal drilling and hydraulic fracturing, which has brought down the cost of new sources of natural gas.

And then finally, oil, where the impacts are going to be muted in the short term, because there are limited substitutes for liquid fuels in the transportation sector, which means the marginal abatement costs are relatively high. So, a cost-effective portfolio in the short term actually would not target oil, but we will see increasing penetration of EVs, growth of biofuels, and greater fuel efficiency. Petrol demand may decline post 2026, but that will be muted by growing demand for aviation fuels and for petrochemicals.

Now, the economic impacts are, also can be looked at as bad news for fossil fuels, but obviously good news for renewables and possibly for nuclear power. That depends upon domestic political circumstances.

In other sectors, climate policies will increase energy costs. So, a simple rule of thumb is that it will be bad news for sectors that use energy, but that is all sectors. On the other hand, it can be good news for the producers of energy consuming durable goods, such as manufacturers of commercial aircraft, because when the price of jet fuel goes up, then there is a more rapid turnover of the capital stock of aircraft, because each generation is exogenously more efficient than the last. So it will be good news for Boeing and Airbus, but particularly bad news for the consumers of those same energy consuming durable goods, whether it is Air India, Lufthansa, or United Airlines.

Finally, I want to remind you in closing, that it is a global commons problem, so international cooperation is necessary. That is why the annual negotiations under the United Nations remain very important. Even though many of us in the room find them frustrating, they are important. But it is also why climate change merits continuous attention from the G 20, which is why this conference over today and tomorrow is, I believe, so important.

So, in addition to saying thank you, I just want to leave you with some websites where you can get more information. The Harvard Project on Climate Agreements, the Harvard Environmental Economics Program, each of which I direct; also my website, my blog, and of course you can follow me on Twitter.

1.3 Speaker 2: Jessica Seddon

India faces the triple challenge of (i) continuing to improve living standards for more of its population, (ii) within a narrowing carbon window, and (iii) in the face of accelerating environmental change. It has to build the societal capacity to do new things—from crop-switching to battery innovation to resilient, low-carbon infrastructure—quickly. In short, it has to become more flexible. This paper sketches a roadmap for a National Flexibility Mission.

The triple challenge. Since the other presentations in this panel have focused on the mitigation challenge, I am going to spend a little bit of time putting a spotlight on the new operating environment for development in India.

Temperatures in the region are rising, albeit at slightly lower rates than the global average.[1] These start from a relative warm base temperature, however, so the frequency, duration, and geographic scale of dangerous heatwaves is on the rise, particularly across the highly populated Indo-Gangetic plain and central India.

Turning to water, which not only supports agriculture, industry, and household needs but also a large portion of the electricity sector (thermal and hydropower): availability is evolving. Monsoon patterns are shifting and the groundwater stocks

[1] Krishnan et al. (2020) find that land surface temperatures of India rose by 0.7C from 1901–2018, compared to a world average increase of 1C.

that could, with further investment in water management and irrigation infrastructure, serve as a buffer for more variable rainfall are depleted. The glaciers in the Himalayas, which act as water storage for surrounding nations, are shrinking.

As far as weather goes: the oceans around India are warming faster than the global average, leading to more rapid intensification of cyclones and coast storms as well as new seasonal patterns of storms. India's economic geography includes significant coastal value-at-risk.

The key point here: environmental change is altering India's operating environment for development.

These risks are in some sense known to the Indian policy world. The striking Fig. 1.1 on temperature increases comes from the Indian Meteorological Department's "Statement on Climate of India during 2022."[2] Ministry of Earth Sciences' 2020 assessment of climate change over the Indian region is a comprehensive and detailed combination of regional and global research on environmental change. Figure 1.2 as well as Table 1.5 and Chapter 11 in the report summarize this succinctly for those seeking a quick read.[3] The World Meteorological Organization's report on ocean temperatures is included in online training packets for the Indian Administrative Services (IAS) exam.[4]

At the most basic level, these changes, along with the global need for India to pursue low-carbon development, are a call to strengthen the overall capacity to do things differently—fast. The list of things that will need to be done differently in a warming, decarbonizing world is long, as is the range of entities and individuals who will need to actually *do* this doing. Farmers will have to grow new crops, governments at all levels will have to make different choices about transport, power, and water infrastructure, and businesses will have to build in new buffers and adapt their value chains to delivering and accounting for new products and services. Workers will have to learn new skills midlife, and households may need to move to more climate-proof areas.

The "push" for change is becoming clearer, whether it is a change in the pattern of the monsoon that leaves some areas unexpectedly dry and others underwater with floods or a cross-border carbon adjustment mechanism that makes carbon intensity more expensive than it used to be.

---

[2] Available at: https://mausam.imd.gov.in/Forecast/marquee_data/Statement_climate_of_india_2022_final.pdf.

[3] Table 1.5: Krishnan, R. et al. (2020). Introduction to Climate Change Over the Indian Region. In: Krishnan, R., Sanjay, J., Gnanaseelan, C., Mujumdar, M., Kulkarni, A., Chakraborty, S. (eds) Assessment of Climate Change over the Indian Region. Springer, Singapore. https://doi.org/10.1007/978-981-15-4327-2_1.

Chapter 11: Dhara, C., Krishnan, R., Niyogi, D. (2020). Possible Climate Change Impacts and Policy-Relevant Messages. In: Krishnan, R., Sanjay, J., Gnanaseelan, C., Mujumdar, M., Kulkarni, A., Chakraborty, S. (eds) Assessment of Climate Change over the Indian Region. Springer, Singapore. https://doi.org/10.1007/978-981-15-4327-2_12.

[4] See for example, in Drishti: https://www.drishtiias.com/daily-updates/daily-news-analysis/global-sea-level-rise-and-implications-wmo#:~:text=Rate%20of%20Sea%20Level%20Rise,inland%20by%20about%2017%20meters.

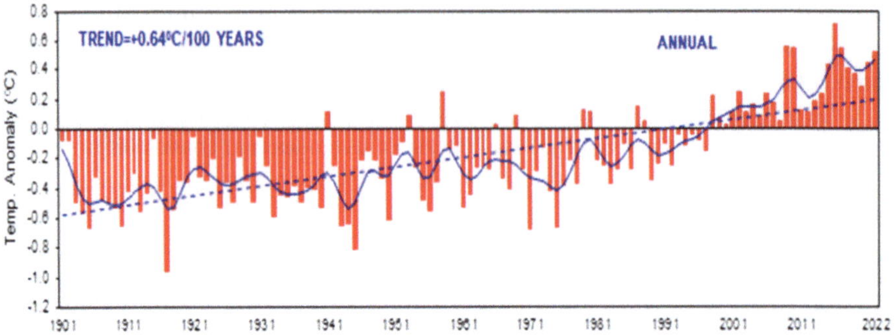

Fig. 1.1 Annual mean land surface air temperature anomalies averaged over India for the period 1901–2022. The anomalies were computed with respect to the base period of 1981–2010. The dotted line indicates the linear trend in the time series. The solid blue curve represents the sub-decadal time scale variation smoothed with a binomial filter

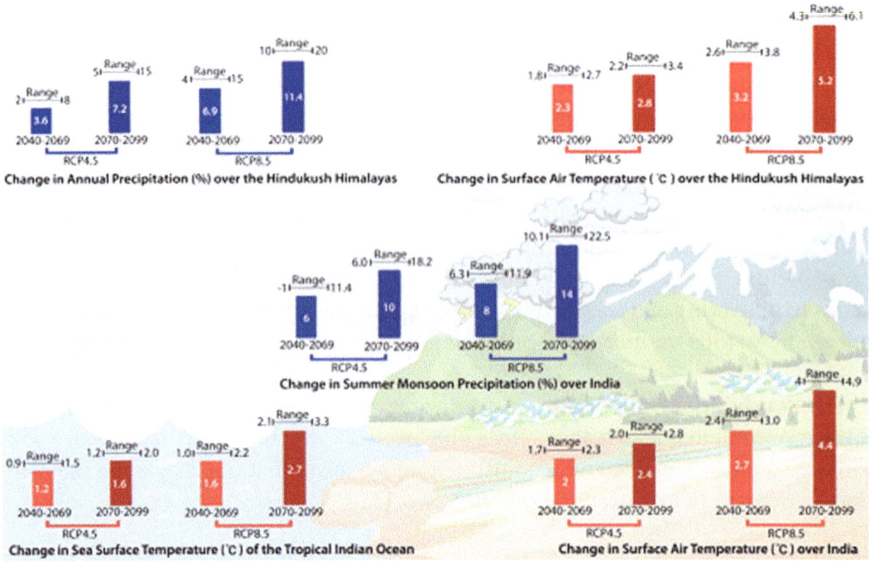

Fig. 1.2 Projected changes over the Indian region

Reducing the frictions—building flexibility—is essential to ensure that this push does not become damaging pressure on lives, livelihoods, and the economy. The case for flexibility can be made in various ways and varies across dimensions. In macroeconomic terms, within the world of climate adaptation, for example, economist Esteban Rossi-Hansberg and others have found that the flexibility to migrate or relocate investments brings the expected cost of sea level rise down from 4.5% of real global GDP in 2200 to 0.11%.[5] Similar cases have been made for the value of workers' ability to shift across firms and industries as technology evolves. In both cases it is important to note that such "flexibility" can be hard for those who relocating or finding new jobs—part of the challenge in building flexibility is making sure that the burdens of flexing are not disproportionately placed on those with less economic or political power. In micro-economic terms, the value of flexibility can be expressed in terms of expanding options for individuals. An individual with a portable identity, a strong educational foundation, access to a safety net independent of employment, and access to credit is far more "flexible"—able to do new things fast—than an individual who is locked into a particular place and job to survive.

How might India build its capacity for flexibility? What might be the formation of an initiative, a set of actions, a mission-level orienting framework for India to essentially be a pioneer in leading in both adaptation as well as mitigation?

Here I sketch a high-level roadmap for a flexibility mission that a national supra-ministerial entity such as the NITI Aayog could lead. The roadmap draws on my academic past in game theory, my professional present focus on distilling workable principles for institutional design, and some of what I've learned over the past couple of decades of having the privilege of working with partners in Indian business, government, and civil society on infrastructure and urban policies. There is not a comprehensive precedent to learn from. While most blueprints for new initiatives tends to start with a recitation of case studies from other countries—in this case, India has an opportunity for being the first to create a whole-of-government-plus-partners initiative toward building the twenty-first-century societal capacities behind mitigation, adaptation, and more.

Orient*—name and describe "flexibility" as a distinct shared goal* for various levels of government, sectoral agencies, business associations, and civil society groups to work toward.

Prioritizing flexibility as a social capability offers an alternative to the "incentives" approach to change. If one is looking at sharpening incentives for a transition, one might use taxes, subsidies, regulatory carrots and sticks, or carbon pricing to create a push. If one wants to instead reduce the frictions to change—the reason that industrial clusters are not moving to gas rather than biomass, the reason that people are not moving from one farming region suffering from drought to a new one—one might look at supporting individual flexible, investing in infrastructure that lowers the cost of switching, or supporting technology and business models that generate real options. Similarly, with infrastructure programs, a "push" mentality may seek to promote a particular form, green hydrogen, a particular industry, a particular special economic zone. A "flexibility" approach might, in contrast, focus on functional, interconnected, multimodal transport as a platform for new choices. That contrast is quite valuable in defining what would be a comprehensive, whole-of-government effort to build the social capability to address both of these.

---

[5] Desmet, Klaus, Robert E. Kopp, Scott A. Kulp, Dávid Krisztián Nagy, Michael Oppenheimer, Esteban Rossi-Hansberg, and Benjamin H. Strauss. 2021. "Evaluating the Economic Cost of Coastal Flooding." *American Economic Journal: Macroeconomics*, 13(2): 444–86.

The initial naming and valuation of flexibility is important because India is a complex governance environment with three levels of government, more specifically delineated sectoral ministries and state agencies than most of its peer nations, a number of different business associations representing diverse firm sizes and interests, and varied economic and cultural geography. Achieving durable, real, change in such a setting requires approaches informed by system leadership. Within system leadership, setting a guiding "north star" that can orient various parts of the system toward contributing to a shared future is the first step.[6]

Diagnose and focus—identify the essential inflexibilities. Focusing on flexibility may be distinct from a more conventional emphasis on using policy to change motivation – but it is still a broad and vague term. The first step is focusing on the forms of flexibility that will be most needed for decarbonization and adaptation. Is it switching jobs and investments across sectors? Is it input switching, for example, in green construction or energy sources for industry? Value chain rearrangement that lessons the transport needs and thus emissions, or enables greater circularity of resource use?

The various forms of flexibility also have their own political economy and other considerations. Migration, for example, is an important form of flexibility, but there are other considerations around economic and social geography as well as personal preferences that may weigh against a focus on this kind of mobility.

After the flexibility priorities have been identified, diagnostic effort needs to shift to hone in on the micro-constraints to change. Why are people *not* making the choices that would put them in a safer, lower carbon setting and profession? What are the reasons that firms are not replacing their energy sources? If "net zero is net positive," as a previous speaker has noted, why is it not happening? Is it access to credit? As a matter of entrepreneurship, what is stopping small and large businesses from investing in these new technologies? Is it access to logistics and transport? Insurance? Working capital? Skilled labor? What is stopping individuals from shifting into new careers? Skills? Savings to support time out of the labor force? Social expectations?

Methodologically, the diagnostic phase could comprise a series of root cause analysis studies of the key changes on the mitigation and adaptation to do list. These cases would inevitably generate a seemingly random, rambling, and potentially very uncomfortable list of frictions to address. Identifying and responding to the underlying frictions to change, however, is an essential precursor to be able to address both mitigation as well as adaptation needs.

Implement—organize, Assign, and Reward the Progress. The list of inflexibilities will inevitably span issue areas and levels of government, but for implementation would need to be clustered into sub-areas of work that can be assigned to (and funded) as part of various agencies' missions. This is often the step at which integrated strategic efforts, from urban planning to climate missions fall apart. Each of the ministries, levels of government, businesses, and other actors has their own

[6] Senge, P., Hamilton, H., & Kania, J. (2014). The Dawn of System Leadership. *Stanford Social Innovation Review, 13*(1), 27–33. https://doi.org/10.48558/YTE7-XT62.

existing set of goals and operating environment and this additional task runs the risk of just joining the queue.

One way to avoid this "valley of death"—to borrow a term from other areas of innovation is to group the first few flexibility investments to be relatively small changes in the emphasis of existing agencies and policy instruments rather than entirely new programs. If adult education and reskilling turns out to be an important area for increasing flexibility, build on the National Skill Development Mission. If access to credit turns out to be a significant stumbling block for industries to switch fuels or new businesses to enter with more resilient business models, expand existing programs for entrepreneurs and small and medium enterprises.

Another way is to encourage flexibility investments to come from within the existing structures rather than impose them as an additional mandate. Reward the leaders across levels and sectors of government, business, and civil society for their contribution to flexibility. There is precedent for evaluating policy changes in terms of their contribution to flexibility. A number of papers, for example, evaluate U.S. healthcare reforms and proposals, for example, in terms of their impact on "job lock," or worker choices to stay in jobs longer than they otherwise would have.[7]

A national mission on flexibility would be politically challenging in several ways. A public focus on flexibility could be seen as an imposition rather than an opportunity—a failure to protect the aspects of peoples' lives that they value. It would be important to build in some way of distributing the burden of flexibility. The initial diagnosis of inflexibility could be politically embarrassing—a litany of oversights and omissions. The efforts to enlist cooperation across sectors and levels of government could run into turf wars.

It would also, however, be a demonstration of globally relevant leadership in building the institutional architecture for new societal capabilities that many countries need to thrive in times of accelerating change.

1.4 Speaker 3: Arunabha Ghosh

Since Professor Stavins has brought out the overarching framework, I thought I would delve deeper into some of the concrete trade-offs. As Mr. Bery has said very clearly, as well as of course, Mr. Jayant Sinha, it's not just that it is net positive in the models, but how do we make it happen and how do we deal with the trade-offs?

So that's really the theme of what I'm going to touch upon. At CEEW, our work is around the system-level transformation modeling, but also trying to translate all that

[7] The majority of Americans rely on employer-sponsored health insurance to address the potentially high costs of health care. Individually purchased insurance is available, but more expensive. For a summary of one such policy evaluation of policies that might change the cost of access to health insurance and thus health care, see Government Accountability Office memo on "Health Care Coverage: Job Lock and the Potential Impact of the Patient Protection and Affordable Care Act," https://www.gao.gov/assets/gao-12-166r.pdf.

into the quality of life issues that ordinary citizens face, supported by the enablers of finance, technology, circular economy, and resilience.

Let me start with the first trade-off regarding internalizing climate risks. The climate emergency is exacerbating fiscal challenges. Let me illustrate this with a couple of examples. At CEEW, we've been developing the first high-resolution climate risk atlas for India. As those maps show, already three-quarters of our districts are hotspots for extreme climate events. 80% of Indians are living in areas that are highly vulnerable to extreme climate events. But what is even more worrying is, if you see the second map, that 40% of our districts are showing trends that are swapping trends. Basically, what was traditionally flood-prone is becoming drought-prone and vice versa. This then challenges not just climate models, in the sense that the past is no longer a predictor of the future, but also challenges administrative capacity on the ground. If a district administration has historically been more used to dealing with droughts, how do they overnight become more used to dealing with floods?

Now, if you translate that at a global level, one of the things that you see, and I want to draw your attention particularly to the bottom panel there, is that the low-income economies and the low-middle-income economies are the ones that suffer the most in terms of the percentage of GDP that gets impacted by climate-related disasters. That's data for about two decades. And you can see for those two categories, the share of GDP getting impacted is well over 1%. Whereas for the high-income or upper-middle-income economies, the share is lower. So this is what is constraining the fiscal space that these economies are already struggling with and which is part of the G20 agenda.

So how is the G20 dealing with it? One of the innovations this year that has been introduced is this new working group on disaster risk reduction with a range of objectives: disaster-resilient infrastructure financing, mainstreaming of disaster resilience into policy, and so forth. Equally, there is a global organization that India is promoting called the Coalition for Disaster Resilient Infrastructure, which also has a focus on specific foundational sectors like telecoms, transport, particularly the airports, power, finance, etc. How do you make these elements of your economy, of your broader economy, more resilient?

How do we deal with that shrinking fiscal space? Last month, at the Paris Financing Summit, for which I was an advisor to the French presidency, we brought out a range of different ideas and suggestions of what could be done. Of course, the IMF has announced a Resilience and Sustainability Trust, leading up to about \$60 billion. Some estimates suggest that the so-called \$100 billion promise might be delivered this year from developed countries and so forth. But equally, we have to be innovative. For instance, illicit financing flows: Africa alone lost about \$88 billion in tax revenue from illicit flows between 2015 and 2020. We've got to act on the green development agenda while not forgetting the broader macroeconomic space creation agenda that is necessary.

Let me now move to the second trade-off. And I say these are trade-offs, notwithstanding what Mr. Sinha said, with which I fully agree, that if you actually do the modeling, the trade-offs might not be there. But these are, at least in perception and sometimes in very real terms, political trade-offs that we have to confront. One of the trade-offs is between energy access and clean energy. Now, energy access is going to be the primary driver for many developing countries as a political priority. Between 2000 and now, India has given access to electricity to 700 million people. Between now and 2030, the world has to give access to electricity to 700 million people, many of whom are in sub-Saharan Africa. Effectively, if India, since 2017 when we introduced the Saubhagya scheme of household electrification, connected 11,000 Indians to electricity every hour over 18 months, 28 million homes got connected. Now, between now and 2030, for the rest of the world, every hour we have to connect 11 and a half thousand human beings to electricity.

So this is the scale of the challenge, which is why then political leaders or policymakers might say, "Look, my priority is to get power to the people." But is there really a trade-off? So at CEEW, we did the first model of looking at net zero that went beyond 2050 while trying to be consistent with planetary integrity.

And what we found in the lead up to Glasgow was that a 2070 target for India still ensures that our emissions are 59% lower than China's, 58% lower than the United States, and 49% lower than the European Union. But it gives us more time. So the real question is not just whether, in the models, net zero becomes net positive, but how do you create that policy room to peak and then bring down your emissions?

Now, if you translate that globally, what we find; based on some analysis I did for all emerging markets in 2021, we see that 88% of all new energy demand over the next two decades will come from emerging markets, but their political economies will be different. Some, like India and China, will have to leapfrog from coal, as Mr. Sinha was describing, and some might have to leapfrog from gas and elsewhere. But that then raises this question of how will all of this happen? Earlier this month, I published a paper in Foreign Affairs titled "Can India Become a Green Superpower?" Now, what does that superpower really mean? Over the last decade, you can see the exponential increase in renewable energy capacity in India. We are already the fourth-largest renewable energy capacity in the world, with about 43% of our installed generation capacity coming from non-fossil sources. But that's not enough. We are also the second-largest national target for green hydrogen, just short of the United States and just short of the European Union's collective target for green hydrogen.

At CEEW, we also have a real-time electric vehicle sales dashboard. You can get that information down to every single regional transport office, and you can see the exponential rise in sales of EVs. However, despite powering sustainability, India, the fourth-largest renewable energy market in the world, is getting less than 2.9% of global clean energy investment. And then, if you compare it with other countries, if India is getting that much, all of Africa is getting even less, Brazil is getting less, and so forth. And this is why the point that Mr. Sherpa made is not just about the Green Development Pact in terms of the policies, but how do we also get to the financing?

The basic problem with the financing is that the risks perceived for investing in emerging markets like India and elsewhere are far higher than the risks realized. How do you actually de-risk? When I analyzed more than two dozen financial initiatives that have been launched over the last decade, less than 10 of them even attempted to tackle investment risks. And this is why the World Bank Reform agenda needs to think about what is the platform for blended finance to look at risks across projects because often these are non-project risks. These are related to currency, policy risks, political risks, etc. So our proposal is what is called the Global Clean Investment Risk Mitigation Mechanism, a digital platform to pool the risks across geographies and across projects to lower the risk curve, and then use the limited amount of public financing as the first loss to bring down the cost of finance.

Let me now turn to the third trade-off, which is around energy security versus energy sustainability. Now, of course, the Indian G20 presidency is happening against the backdrop of serious concerns about energy security, but let me also argue that it is not just energy security of the fuels of the past but also energy security of the fuels of the future that we have to be concerned about. Take India again as the starting example. If you look at the right-hand panel, our fossil fuel countries that export fossil fuels to us are fairly diversified. The regions from where we get our renewable energy equipment are much more concentrated. So, taking it beyond India, the G20 tasked us with developing an official study on renewable energy supply chains. And you can see over the past decade, as many more countries have started investing in clean energy, the dependency on concentrated imports has gone up: in solar, less than 40 countries with concentrated sources of renewable energy imports going up to over 70; in wind, it has kind of stayed the same; lithium and batteries have gone up from 19 to 49. So energy security for the fuels of the future is going to matter to make the political case that the investments in clean energy can also be consistent with energy sustainability and energy security.

So, just last weekend at the Goa Energy Transition Working Group ministerial, we, of course, did not have a joint communique for reasons we all know. But in the outcomes document, we see some very encouraging signs. For the first time, there is an extensive paragraph on critical minerals for batteries as well as a separate section on fuels of the future, along with an agreed high-level principles on hydrogen, as well as of course, the usual mentions of bridging technology gaps and universal energy access and so forth. So we need to now leverage these agreements.

Now, let me finally come to the very last bit, the last trade-off, around job growth and job losses. Again, to pick up on the theme that Mr. Sinha very clearly articulated, and not just his constituency, but if you look across the country, we've got 266 of our 700-odd districts with at least one asset linked to coal, 13 million people formally employed, and there's a much larger informal sector as well. Now, can we combat this? At CEEW, we collect data on the number of jobs being created in the clean energy sectors. We estimate that by 2030 we'll have a workforce of over a million people in large-scale solar and wind. But that will translate to nearly three and a half million full-time equivalent jobs. Beyond that, if you look at the role of distributed energy to power livelihoods, especially in rural areas, our estimation is that it is a $50 billion investment opportunity with the potential to support 37 million livelihoods. And I can get into more details on that front. We also did analysis on the job gains and losses in the electric vehicles versus the Internal Combustion Engine (ICE) mobility sector.

And you can see that there is a net job loss if you move away from ICE. However, you can combat it by creating new industries like battery recycling, installation, or charging infrastructure, and so forth. So there are ways to overcome even that major political conundrum of job growth or job losses.

So, as I conclude, I want to simply bring it back to what this means for, as Professor Stavins said, what need not be global, but at least international or multilateral of some kind. We've got to solve for four market failures and four political failures through the G20 and beyond.

The first market failure: going back to the issue of risk, how do we deal with non-linear climate risks that are rising over time? How do we create an insurance cushion through a resilience reserve fund? How do we deal with that delta between perceived and real risk by creating a de-risking platform? How do we price externalities, not just carbon, but land, water, materials? By promoting a circular economy that India has put forward in its G20 agenda. And how do we promote sustainable consumption, not just the supply side of sustainable production, through "Lifestyle for Environment," the high-level principles for which were agreed at the Development Working Group last month.

Equally, there are political failures, and as much as we are frustrated with the core processes, the challenges that we have a lack of accountability, not a lack of promises. How do you convert the UNFCCC from a bank of depositing promises to a bank of actions? How do we move away from protectionism, from concentrated supply chains to more diversified yet interdependent supply chains through technology co-development. How do we have security for the fuels of the future through the rules that will govern these new fuels? And finally, going back to the issue of jobs, how do we have an orderly transition from fossil fuels through not just energy transition partnerships but joint energy transition partnerships?

I leave you with five questions. I think about these for India, but perhaps they apply to the rest of the world as well. What are the macro enablers for this domestic green policy? How are we going to increase the fiscal space to deal with climate risk? Are we politically ready to price resources? Can we manage the macrofundamentals of inflation or currency variability, which increases the cost of finance for clean investments? Do we have the commercial diplomacy to support our strategic diplomacy, supported with trade and industry policy that creates these interdependent supply chains? And finally, do we have the political maturity within and across our countries to handle the distributional consequences?

Chapter 2
Technology, Policy, Jobs

Sachin Chaturvedi, Paul Samson, Albert Van Jaarsveld, Debjani Ghosh, and Vijay Kumar Saraswat

2.1 Session Chair: Sachin Chaturvedi

I think it's absolutely important in terms of thinking about the new development model that we need when we are talking about several challenges that are there. I think, as all of you would agree, the technology, positioning of technology and the policy that we require for technology, it works in an ecosystem.

And that ecosystem requires us as in the previous session, we heard Jessica and Arunabha both alluding to the idea of industrial policy. So how technology policy, industrial policy, and trade policy converge or diverge in what way incentives or disincentives come up? The act that was referred to in the previous discussion in terms of how the US subsidy regime is actually accelerating the quantum that is required to support and also to absorb the kind of changes that we are looking at.

So, from that perspective, the role that multilateral institutions should play, the kind of positioning that they can do, particularly in an era where we are seeing a

S. Chaturvedi (✉)
Director General, Research and Information System for Developing Countries (RIS), New Delhi, India

P. Samson
President, Center for International Governance Innovation (CIGI), Waterloo, Canada

NITI Aayog Conference On Promoting Green and Sustainable Growth, ITC Maurya, New Delhi, India

A. Van Jaarsveld
Director General, International Institute for Applied Systems Analysis (IIASA), Laxenburg, Austria

D. Ghosh
President, National Association of Software & Services Companies (NASSCOM), New Delhi, India

V. K. Saraswat
Member, NITI Aayog, Delhi, India

S. Bery et al. (eds.), *Navigating Challenges for Sustainable Growth*,
https://doi.org/10.1007/978-981-97-7894-2_2

transition in the very nature of jobs that are being created. If you pick up the latest report from UNIDO on the Asian scenario in industrial policymaking.

The amount of effort that all Asian countries are putting together is absolutely clear. They all are going in the direction of evolving industrial policies. They are evolving mechanisms which are strengthening the processes. And they're also looking into largely the implications of digital labour platforms that are coming in and how they are transforming the future of work, in what way new models of development they are contributing to.

And also, in terms of new labour relations that are coming up along with that. And in that process, if we see the crisis of multilateralism, which gets juxtaposed in terms of adding more to that complexity that we are talking of the impact of exogenous shocks. We talked about climate change, but then certainly the debt crisis, the financial crisis, sometimes the national challenges that are there, I think they all in a way complicate the situation and the scenario where technology has to perform and the nature of technology that is to be absorbed.

The support mechanisms that these institutions bring in also require a bit of a challenge even within India, if you see a large amount of effort has gone in last couple of years to enhance the size of the digital economy in India. In 2017–18, when it was somewhere close to \$200 billion. We are expecting by 2025, it would be \$1 trillion. And this would be something which would be huge in terms of how we look at in terms of the job profile that would come up.

Even the data center economy has expanded, which was something like \$4.4 billion in 2021, and we are expecting that by 24–25, it would be close to \$8 billion. So, you can see the pace at which this change is happening and the amount of change in the nature of jobs, the nature of insecurities that are coming in.

The way labour unions, the way groupings on labour are looking into. So, the standard design that IMF and World Bank follow at some point in terms of structural adjustment program, we need to see what way we bring in a large chunk of people within this. And that largely, as chair very rightly pointed out, brings in the new policy model that is needed. And that also in a way contributes to the larger area of coping up with several exogenous shocks that are coming in.

We have an excellent panel today that is going to bring in several of these dimensions together. Bringing in issues which are related to migration that is happening, issues that are related to Artificial Intelligence (AI) and its role. As you know, in the G20 presidency of India, as was also in the Indonesian presidency.

The AI and ethics, they were analysed, commented upon. And India's own declaration would bring some of these concerns forward. And that would also contribute in terms of how the wider development policy framework comes up.

2.2 Speaker 1: Paul Samson

I am going to stress some issues that I think are first order issues for what we are talking about today. I will say right up front that what I am trying to do is introduce the concept of global change in a broad sense, and then we'll talk about demographics, economies, and then move on to the environment and technology.

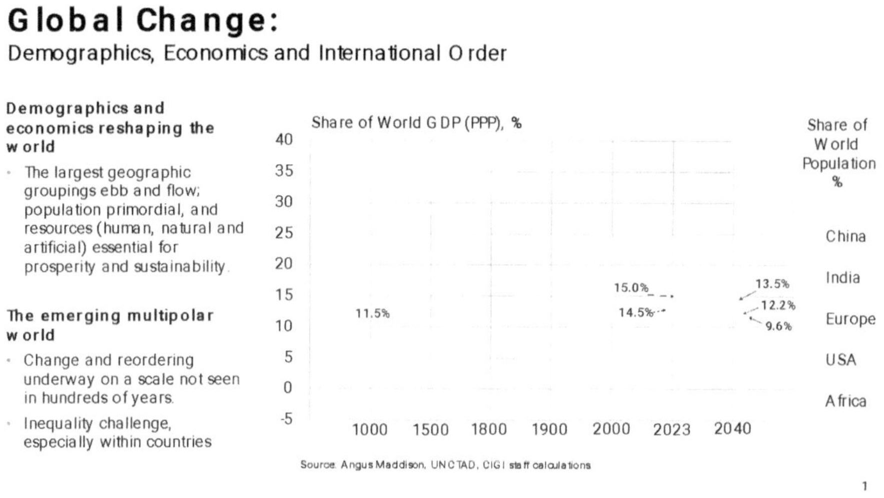

I'll add some caveats right away that this graphic is intentionally provocative. Anything that goes back 1000 years in terms of data can be challenged. This chart has as the vertical axis growth (global percentage of the economy measured in purchasing power parity (PPP)) and time on the horizontal axis, over the last thousand years. The size of the bubble is the share of the world's population. The caveats are, of course, that the grouping of these countries can be challenged. The groupings used here have generally included more rather than less. For example, it has grouped Africa all together, which includes North Africa and Sub-Saharan Africa.

The point in this slide is to emphasize a couple of things. One, is that demography tends to be the great equalizer over time. That the large blocks or groupings have an advantage through the size of their population to generate more internal economy, and to become a bigger global economic force.

But at the same time, you can see the emergence of technology playing strongly here. Other historical factors matter too. And we could spend a lot of time on this, but what I really want to point you towards is what happened in the last 20 years or so, where you have a couple of stories. One is that if you look at the 2000 data there, which is the third from the right, you see an incredible story of China's economic emergence, or re-emergence. We can debate whether PPP is the right way to measure the economy or not, but the trend is still there. An incredible jump in 20 years. And at the same time, you have, now projected an incredible jump for India over the next 20 years. But in the case of India, both a relatively young population and a rapidly growing economy. The United States interestingly is relatively stable in terms of a share of the global economy from now to projections for 2040.

The other thing I wanted to point out here was Africa, which again is all of Africa. It is emerging rapidly towards a huge share of global population, and by 2040, is the most populous area in the chart. And so a key question is, what will the economy do there?

My next point is on exponential change, and it's come up in many ways in the conversation already. Humans are not very good at thinking about exponential change. Institutions are not very good at adjusting to exponential change. And yet if you look at what's happened over the last, hundred years, couple of hundred years, and certainly the last couple of decades, many things are changing exponentially.

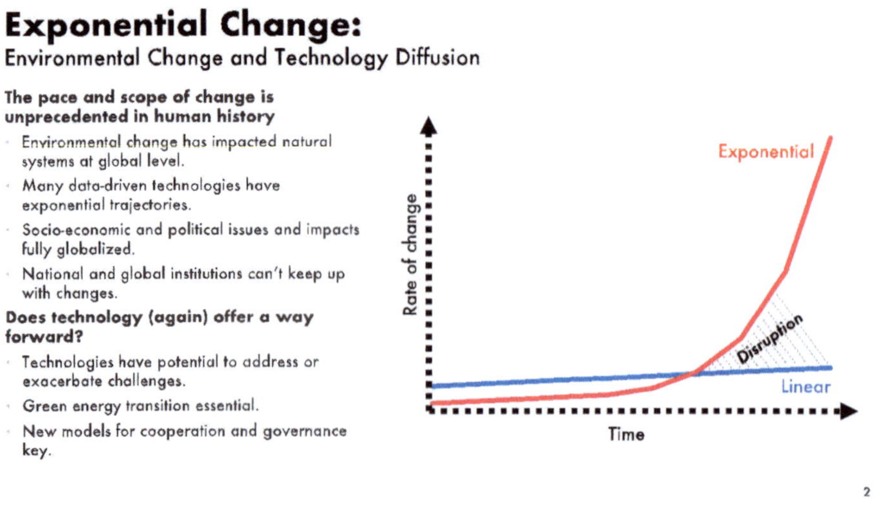

When you think of this concretely, of course, linear change adding one increment, whereas exponential change can be doubling or more. And suddenly you've blasted off in a way that has not been forecasted; or at least the change has not been absorbed by institutions.

What does this tell us about where things are going? I would say that, on the environmental side it's quite easy to look at what has happened since the industrial revolution. This chart shows CO_2 concentrations, both from ice core samples and from real time samples in the atmosphere now, and some modeling.

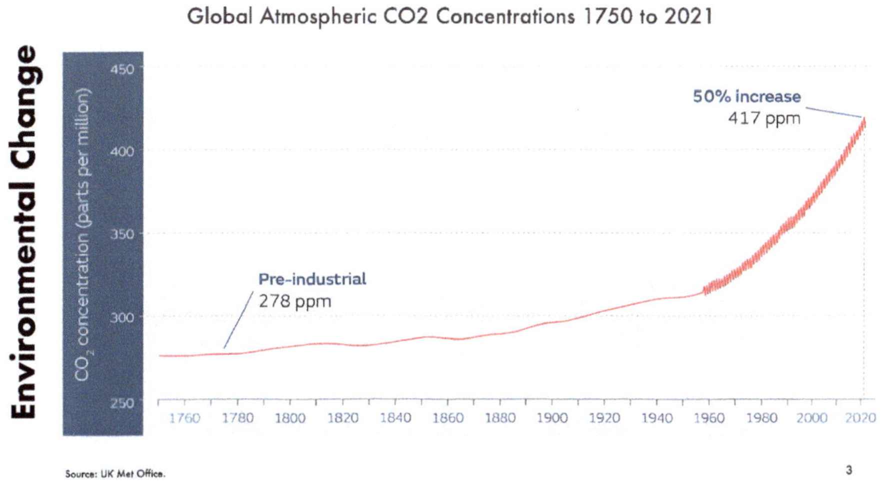

CO_2 concentrations are clearly a case of exponential change. It was already growing quite rapidly due to human population, but in a relatively linear manner, and then it exploded. And I think you could draw out the same parallel for any number of environmental issues. You could do the same thing for biodiversity loss, similar things for oceanic pollution, forestry, and other natural systems.

One of the things that the last slide was trying to emphasize is that the transition from linear to exponential can create higher uncertainty; higher chances of risk; and a higher chance of surprise. Those dynamics are playing out right now on the environment side. I was very glad that climate change was both presented as a mitigation and an adaptation challenge in the previous session. But there is also the risk of surprise in their tipping points occurring; a black swan event that when you look at backwards seems relatively predictable—like the North Atlantic current scientists have been concerned about for decades.

On the technology diffusion and adoption side, again, the linear trends changed over much of human history into exponential trends. And when you look back over the last hundred years or so, or 150 years, you start to see sharp diffusion curves for most technology. Whether it's running water, refrigerators, these kinds of things start to take off quite sharply. But then in the last couple of decades, the exponentiality has increased much more. And it has increased at a global level that is unprecedented, where you might see it not necessarily uniformly globally, but certainly globally simultaneously.

The examples here are ones that people will be familiar with, that took several years for this very sharp diffusion curve to ChatGPT, Open AI's online tool that reached a level of fusion, not seen since the game Candy Crush, which was irresistible. And so we're seeing this across many, many trends. Certainly on the technology side, I would argue also on environmental side.

Finally, on to artificial intelligence (AI). Now one can debate how much artificial intelligence has impacted innovation and productivity so far. But AI has now gone mainstream. I think the moment, as with the apps cited, something is accessible in a way that has not previously been the case, the diffusion curve takes off. So that's what we've seen with AI. I think that there's no question that it is what is referred to as a general purpose technology, which then tends to spin off all kinds of other things. Examples in history are electricity, probably the printing press, and some would go even further.

Of course there have been booms and busts in the past, so what does this mean right now? I think that countries are struggling with how to integrate AI. I think there is an important context that was mentioned already several times about the new industrial policies/protectionist policies in many cases have made the adjustment to AI that much higher stakes because this is one of the areas which may get away from some of the traditional economic growth engines in recent decades.

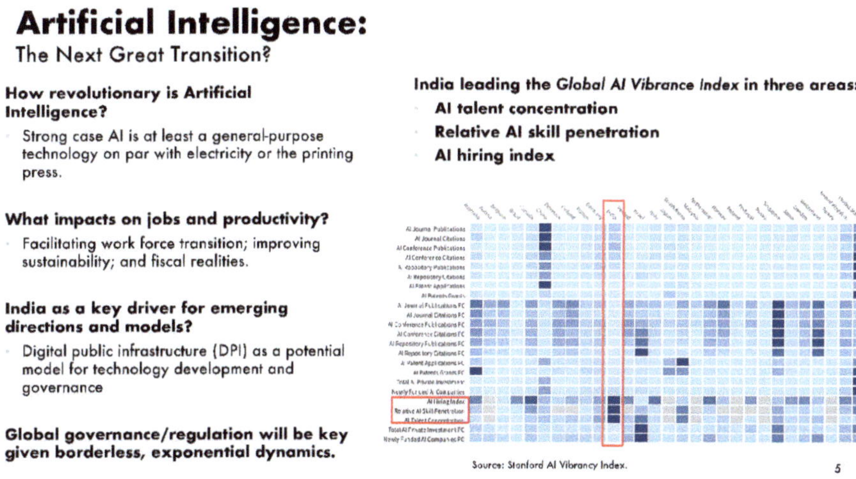

One wonders how automation could be a good thing for a country like India that is seeking to create as many jobs as possible with a young population. But I think that in history, the jobs transition often, as was mentioned earlier today, can work out to be a net positive. The phrase used earlier: net-zero is net-positive overall. You know, AI could be a net-positive overall, I think if managed appropriately and properly like textiles and other mechanizations that have occurred.

What is the path for India here? I think there is a soft power opportunity where India can promote its own interests and demonstrate a path for developing countries. I think the India stack; the digital public infrastructure does propose a very interesting model. I think there are positives on settlements and digital ID and payment systems, and remaining challenges on privacy governance.

I think for a global conversation, there is an opportunity for India to promote that. It is very interesting because you have China and the US as the leaders you would expect on these global effects or AI. But India's trajectory is interesting, and I think other developing countries could also position themselves on hiring AI skills and AI talent that are preparing the economy of the future.

2.3 Speaker 2: Albert Van Jaarsveld

This input will focus on some of the policy and jobs issues that we see from our side as an institute, an institute that prefers to integrate across disciplines and across ministerial mandates to find pathways to a sustainable future.

It has now also become apparent to all that we are not doing great as far as progress with the SDGs are concerned. This problem is a real focus of IIASA work, and we are always exploring ways to advance the 2030 Global Agenda. But first, a quick overview of the executive summary of the World Economic Forum (WEF) Report on the Future of Jobs [1] to highlight some of the key messages:

i. Technology adoption is clearly seen as a key driver for business transformation.
ii. A degree of job creation and destruction from environmental, technology, and economic trends is anticipated.
iii. The technology job impacts are net positive over the next five years.
iv. In the space of the human–machine frontier, it appears that automation is running at a slower pace than originally anticipated.
v. A structural labor market churn of some 23% is anticipated over the next five years with growth in skilled technology and sustainability domains and declines in routine administrative functions. Analytical thinking and creative thinking will be some of the most important skills for workers moving forward.
vi. Many of the report respondents felt positive about their ability to train the capacities they need within the context of their workforce, but
vii. There were some concerns expressed about getting access attracting the right talent to drive industries forward.

This WEF report provides a sense of the anticipated jobs environment over the next 5 years. The focus now shifts to some of the externalities that are likely to impact these job trend expectations over the coming years.

First, is the impact of aging global populations. The proportion of the global population that is over 65 years of age is increasing, and this growth in the elderly is outpacing new population recruits. This trend is anticipated to accelerate into the future and the transition to recognized aging populations will expand. Aging populations have consequences as they affect the proportion of the active workforce, the proportion of taxpayers, and the degree of social dependency for the elderly. This phenomenon continues to raise policy and political concerns across the world. One potential resolution is to try and stimulate labor force participation by interventions that will promote broader participation in the workforce, especially female participation and to try to encourage people to work slightly longer, beyond 65 years of age, to ensure that critical skills are retained across the workforce.

A key driver for acquiring the required future skills in an economy is education. Education also fast-tracks the process of population aging because it affects fertility patterns. It is possible for countries like India, with growing populations, to achieve a demographic transition by 2060 if the right progressive post-secondary and upper secondary education policies and incentives are in place [2]. However, even a focused education drive, can fail if access to basic human needs in terms of physical well-being and safety are not in place.

Rapid transitions in national population demographics are possible. Over the last 60 years, the Republic of Korea has transformed itself from a typical youthful population to a typical aging population [3]. The consequences of this transition are stark and occupying the minds of policy makers in the Republic of Korea at present.

Important IIASA work has demonstrated that even a shifting demographic pattern does not necessarily threaten the prospects of a population being competitive into the future. Here it is important to shift the policy emphasis from the size of the workforce towards the quality of that workforce in terms of its productivity and what it can deliver. Quality, more than quantity, will determine competitiveness at the end of the day. Consequently, there is an opportunity to mitigate the consequences of an aging population at least partially by ensuring that the quality of the workforce increases over time through appropriate education interventions [4].

A more sensitive part of the demographic debate, is of course, the issue of migration. IIASA work has demonstrated that on average about 0,65% of the global population is always on the move, migrating from one part of the world to another. However, this background pattern will likely be altered due to emergent patterns of environmental and climate change [5]. There are some regions where the current migration patterns will increase and other regions where they will likely decline over the coming years due to environmental pressures.

Increasing global inequality will also bring its own pressures to bear. To demonstrate the formidable challenge this problem poses, modeling work conducted at IIASA demonstrates that for just a few Earth system variables (nitrogen, phosphorus, land use, water, and carbon emissions), achieving equitable access for all would require telling and radical shifts across these variables, especially for carbon emissions [6].

Dealing with the rising inequality conundrum may require new and radical socio-economic solutions [7]. One approach, which has been tested with some success (see Table 2.1), and which is becoming a popular global discussion, is the potential value of a basic income grant. Table 2.1 presents a short summary of the potential benefits and negatives associated with a basic income grant. This summary is taken from the Moneycrashers website (see reference below) and includes some enhancements about potential counter arguments, supporting or mitigation stances). It is clear from Table 2.1 that this approach merits serious policy consideration and that many reasons not to implement can be reasonably easily dealt with. The high-cost argument may however eventually be overrun by the scale of social disruption faced by highly unequal societies. Maybe there will be a point where societies can simply not afford to implement a basic income grant.

In determining sustainable development progress, IIASA typically focuses on the broader matter of human well-being rather than only the narrow SDG measures of economic, social, or environmental performance. The IIASA measure of "Years of Good Life" [8] includes quantitative and qualitative assessments of human well-being, including life expectancy, mental well-being, and more subjective measures such as happiness. This overarching measure also gives us a more inclusive picture and systemic view of whether we are making progress in terms of sustainable development as a society [1].

Table 2.1 Basic income grant (BIG) as a potential game changer: This table summarizes reasons to implement or not implement a basic income grant adapted from Moneycrashers—https://www.moneycrashers.com/pros-and-cons-universal-basic-income (Counterarguments or supporting or mitigating responses are provided in brackets and relevant case studies are listed at the end of the table)

10 Good reasons why?	6. Reasons why not?
1. Reduce poverty	1. High cost (may not be able to afford not to implement due to social disruption. Estimates provided thus far are based on First World (rather: industrialized nations') calculations and suggested that level of basic income was either too low or that the cost of provision was too high; BIG can be set at affordable and impactful levels)
2. Reduce inequality	2. Reduce incentive to work—promote laziness (no evidence from Kenya, India, and Finland—BIG should not be too generous)
3. Eliminate the need for government programs	3. Extra money to those that do not need it (BIG can be recovered through tax system from people with income. Major secondary benefit is that everyone that receives BIG is on the tax system)
4. Improve physical and mental health	4. Diminished self-worth—will encourage alienated idle (not relevant if all receive it and should not necessarily be a comfortable living wage. Would avoid paycheck capture and allow people to consider more meaningful careers)
5. Make higher education more accessible	5. Reduce wages (should be lower than a comfortable living wage—minimum wages can be legislated)
6. Improve wages (Alaska—no inflationary impact)	6. Does not address the root causes of poverty—addiction, poor health, lack of education and skills (BIG will increase coping capacity—World Bank)
7. Support care givers	
8. More freedom for domestic violence victims	
9. Encourage entrepreneurship	
10. Protect workers from economic shocks	

(continued)

Table 2.1 (continued)

10 Good reasons why?	6. Reasons why not?
Case studies • The world's first guaranteed income program, called the Speenhamland System, saved many families in rural England from starvation between 1795 and 1834 • Between 2007 and 2009, the Basic Income Grant program in Namibia cut the nation's poverty rate nearly in half • Between 2003 and 2015, a guaranteed income program in Brazil called Bolsa Familia cut that country's poverty rate by more than three-quarters • A 2016 University of Alaska study of the Alaska Permanent Fund, a program that gives a modest cash payment (around $1,000 per year) to all state residents, found that it kept between 15,000 and 25,000 Alaskans out of poverty each year • In the 2010s, UBI trials run by GiveDirectly in Kenya and Uganda boosted participants' earnings, assets, and nutrition • In 2017, a basic income pilot program in Ontario helped participants save more, pay off debt, and improve their living standards • Another trial that same year in Finland significantly improved participants' financial health A 2019 UNICEF report on Iran's basic income program, which gave Iranians monthly cash transfers equal to about $1.50 per day, found that it had significantly reduced poverty in that nation	

A further externality that will likely impact employment and jobs is the increased tendency for economic club formation across the globe [9]. These economic trends, if persistent, will likely further skew migration patterns and people skills exchanges into the future.

Lastly, it is imperative for achieving the SDGs and sustainable development that a more effective multilateral system of cooperation is promoted and developed. The range of geopolitical forces at play today, together with the externalities discussed here, and that can impact on jobs and skills requirements of nation states, suggests that no nation state will resolve their skills, productivity, and sustainability ambitions unilaterally. A sustainable future requires systemic thinking, and effective multilateral cooperation around matters of mutual interest to develop impactful developmental pathways. In this regard IIASA has made inputs into the G20 process by convening policy discussions in areas of reform required in the UN system, the World Trade Organization, the World Health Organization, as well as climate finance environments. It is our hope these conversations will help ensure that the multilateral system becomes stronger in the future, rather than being weakened by many current geopolitical trends.

2.4 Speaker 3: Debjani Ghosh

I focus a bit on the policy dilemma. Actually, when building the presentation, I thought I'd call it "policy imperatives." But we do not know today what the imperatives are. We have opinions, we have a lot of opinions, and even more questions. But I think the next few months are going to be critical for us to get together and figure out what the imperatives are and change it from the dilemma to imperatives and answers.

So, I'm going to cover three broad things. I'm going to talk very quickly because Paul has touched on some of it: the step change in AI and the key considerations. Why should we care about it? Why should we care, and what are the key policy considerations that come out of it?

We've been talking about AI for a very long time. At least since 2015 or 2016, we have been saying AI is going to change the world. Around 2021, we shifted to saying cryptocurrency is going to change the world. Then, in 2022, we again said AI is going to change the world. But something happened in the last eight or nine months. In November 2022, OpenAI launched an application on top of a GPT, a generative pre-trained transformer they were building, called ChatGPT. What we saw is the beginning of the race to AGI. That's what the industry calls it. Suddenly, we saw a step change in technology. It completely changed the dynamics of how we humans relate to this technology called AI. AI has been in our lives for a long time. Every time you use your phone or get into an automated lift, AI is there, working in the background. But suddenly, it became a part of our lives. It became a co-pilot for us. That was the real big change we saw.

I want to talk very quickly about some of the key considerations it's going to bring about. First, the ushering in of the era of generative AI, where we have technology that is mimicking human intelligence. While there's a lot of excitement about the images created by AI, the real magic is its ability to understand language and the context behind language, which in my mind is game-changing. It integrates a layer of intelligence into everything we do.

Four key considerations I want to leave you with:

1. Is this the next big platform shift? We believe yes. The first was the internet, then came the cloud. In 10 years, how much penetration did each achieve? Consumer penetration cloud was 31%. US smartphone was 54.79%. We believe that generative AI will cross 65% penetration, which seems conservative given the pace of adoption we see today.
2. The importance of language in generative AI. AI today, thanks to its ability to understand context and human language, is nearing human baseline across modalities: text, speech, and images. We've seen chatbots pass US law exams and Google's MedPalM2 score 86.5 on MedQA, a tough medical examination. And even more interesting, was that a panel of 15 human doctors? Rated med Palm two versus other human doctors, and they found that out of nine parameters, machine AI did better on eight versus human doctors. The only parameter where it did less than human doctors was accuracy.
3. The shift in technology adoption. It took platforms like ChatGPT just five days to reach 1 million users. The trigger? English. You didn't need to be a data scientist or know any coding language to use AI. English became the most important coding language.
4. Why should we care? A foundational layer of intelligence will be embedded into every product and service. Goldman Sachs said that if implemented right,

> we could see a 7% increase in global GDP. AI is becoming a tool of economic growth, a key driver. If the US implements this right, they could get a 1.5% boost in productivity, essential for a country with a rapidly aging population and declining productivity.

This is going to be immensely important for all countries to think through. Then comes solving for mega challenges, from drug discovery to climate change, then comes national security. In fact, it was interesting to see, last week, the AI team or task force that President Biden has set up. This is what came out of it: experts urged the US to increase competitiveness with China through AI and spectrum. The bottom line was, if the US doesn't lead, China will. This applies to pretty much all countries. If you are thinking about national competitiveness and you're not applying the technology lens or the AI lens, there is a problem.

There are three things that are shaping or driving this evolution that I talked about: access to data (and not just a lot of data but access to a lot of high-quality data), access to AI computing, and access to the mega bucks. Who has that investment? What we are seeing is companies and countries with access to all three are shaping the AI roadmap and impact today. Today, if you look at it, most of the foundational models which are driving or powering the AI revolution are built in the US, and China will soon catch up.

What about other countries? Do we want to have a say in the shaping of the AI roadmap or not? That's the question we need to debate. And how do we regulate everything that comes with it?

For those of you who read Collingridge, what he said is becoming so relevant because if you regulate it early, you're going to stifle innovation. But if you wait till it's mature, it's too late to regulate. So where do you start? Where's the balance? That's the question we need to ask.

The key challenges are the pace of technology, which is evolving way faster than we humans can adapt. Then we have ethical considerations, bias, misinformation, the high cost of development, and the increasing carbon footprint of training the models. This is going to be the opposite of everything we heard in the morning and progress towards the SDG goals. There are also tremendously high entry barriers, job displacement, which both Paul and Albert talked about. And last but not least, when you talk about regulation, one country's regulation will soon become another country's competitive advantage. Given that AI and technology do not know or respect boundaries, if you stop it in one country, it can easily move to another country to continue with the innovation.

I'm going to quickly leave you with six key areas that I think need to be thought through as we consider how to regulate AI:

The balance between innovation and regulation. Do we take the approach of saying, let's not regulate r and d, but let's regulate commercialization? And then of course is the example of what happened with the atom bomb. And with Oppenheimer coming out, it's raising a lot of questions? But, this is something fundamental that has to get debated.

Do we need new laws? What is AI doing? It's increasing our ability to misinform. It's increasing our ability to steal other people's ideas and art. It's to carry on maybe very sophisticated levels of scams. But these are behaviors that have been existing for a very long time.

So, do you need new laws or do you need to update the existing regulatory ecosystem that exists in countries to ensure that they are adapted and they are ready for the world of AI? I talked about how AI doesn't respect any boundaries. So a global approach is going to be absolutely critical, but may the force be with you as you figure out how to bring in all countries to arrive at a global approach.

I think harmonization around design principles in my mind is the best thing that we can do, and I do hope that G 20 kicks off that discussion. Barriers. We've talked about moats. We've talked about entry level barriers. This is important to consider because this will shape who controls AI. Are we going to leave the control to only a few big companies and a few countries?

Or will everyone, or should everyone have a say? It's gonna be very difficult for a country like India to catch up if we follow sequentially. But what India has is tremendous engineering power.

Catching up is gonna be tremendously difficult. Race to AGI, who decides how it'll be used? It is happening. How do you decide, and last but not the least, the role of industry. I just wanna leave you with this. US announced that how industries have come together to create a self-governance framework.

Industry in India, the entire tech industry in India, 30 companies actually led this effort, came together to create a self-governance framework for generative AI, which is available publicly for anyone who's interested. So this is a very important part of thinking about regulation.

2.5 Expert Comment: Vijay Kumar Saraswat

After listening to all the experts who spoke on various subjects related to technology, jobs, and policies, I would like to connect the three in the context of sustainable economic development.

While we all agree that policy encourages the creation of disruptive technologies, technology enables the disruption of jobs and stimulates growth, and jobs, of course, deliver growth and thereby, influence policy. This is the dominant rhetoric of today, with examples like AI and industrial automation in mind. At the same time, we must remember that there exists another virtuous cycle that is often overlooked. Here, jobs incorporate technology to improve growth, and technology amplifies this growth to drive policy, which in turn, creates more such jobs. We should recall that it was this kind of cycle that led India's ICT services sector to become the global leader that it is today.

The main thing which has been discussed so far is that technology has been significantly affecting the job landscape, and the dichotomy presents a unique challenge for policy-making as to how to balance it. We should also look at these technological developments in the context of the forecasts made by very eminent scientists like Stephen Hawking and even tech philanthropists like Bill Gates, who mentioned that generative AI is likely to be one of the occurrences leading to doomsday and is going to have serious impacts as far as Planet Earth is concerned. Notwithstanding the severity of these predictions, we must acknowledge that these disruptive technologies change how individuals live and how society works. We must see it as a process of "constructive demolition" of old ways.

So, when we now make our policies or important decisions, I think some of these forecasts have to be taken into consideration. In that context, governments have introduced several policies aimed at promoting technologies and adoption and mitigating the adverse effects of that. India, of course, has been pursuing its digital economy exponentially over the last few decades with increased penetration of the internet and digital infrastructure development and startup ecosystem.

Today, we have the second-largest online market worldwide. The IT Business Process Management industry is a significant contributor to India's GDP and employment industry. Revenue in India reached about 194 billion in 2020, employing about 4.36 million people. I'm just trying to say that this situation is also because information technology first disrupted and then created jobs. We must understand that the rise of new technologies is simply an exogenous shock, and we as policymakers are responsible for creating opportunities and employment from them.

I acknowledgement that while technology has been a catalyst for job creation, it also poses significant challenges in terms of job displacement due to automation. I will give you some examples of both cases. On one hand, Indian unicorns and startups have created over one lakh jobs every year since 2019, and new roles have been created like data scientists, AI specialists, and digital marketers. But on the other hand, It has been predicted that about 69% of jobs in India are threatened by automation. It is estimated that 56% of the salaried employment in Cambodia, Indonesia, Philippines, Thailand, and Vietnam are at high risk of displacement. In India too, it is estimated that by 2030, 9% of our work hours in India could be automated. So, there is a challenge for policymakers to leverage the job-creating potential of technology while mitigating its adverse impact.

In addition to that, the rapid pace of technological advancement has also brought challenges, especially for traditional sectors like manufacturing and services, which are particularly vulnerable to automation. The rise of AI, as brought out by earlier speakers, and machine learning (ML) has raised concerns about job displacement. Policy plays a major role here in ensuring that these technologies can be effectively catalysed to yield jobs, and in turn, sustainable growth.

In that respect, I think the Government of India has taken major initiatives like our Digital India, which is responsible for empowering every citizen with a full suite of Digital Public Infrastructure (DPI) products and services. Similarly, there has been an effort to leverage and cultivate the innovative entrepreneurial spirit through the Startup India initiative and the Atal Innovation Mission. These policies are certainly very important for leveraging technology to create a future of jobs that is inclusive, sustainable, and prosperous.

Now, certain recommendations should be looked at for driving the job market. These include policies like skill development and reskilling, social security for displaced workers, inclusive digital infrastructure policy, and regulation of emerging technologies like creating an independent regulatory body for AI and ML to address key issues raised by earlier speakers. Public-private partnership is one of the major requirements, and international cooperation will certainly pave a major way.

I also agree with earlier speakers who spoke of a multilateral approach to solving these challenges. Measures like exchanging best practices, collaborating, and coordinating policies on forums like G20, Digital Learning, and Economic Task Force will be important. There is a major advantage of adopting frontier technologies, which help in economic growth, development, job creation, improving public services, addressing social changes, and promoting innovation. It is my firm belief that a focus on such socioeconomic development-oriented applications of disruptive technologies will help ensure they transform, rather than displace, jobs. Which, I believe, will yield long-term sustainable growth.

The Government of India, and more specifically the NITI Aayog, has been actively working towards this goal. This begins with a range of modern policies on cutting-edge technologies like the National Policy on Electronics 2019 and the Indian Space Policy 2023; covering crucial missions like the National Supercomputing Mission and the National Quantum Technologies Mission; and extends to strategies for emerging technology areas like the National Strategy on AI and Blockchain and the 6G Vision. So there should be some adoption of these frontier technologies which will help us in solving the problems which we're discussing today. For that, we should develop a comprehensive policy framework for the adoption and regulation of frontier technology.

We should invest in building robust and inclusive digital infrastructure. We should improve digital literacy, which will bring in more skills and jobs. We should bring in public-private partnerships and create more international forums to learn from global best practices. The Industry should look at invest in R&D in frontier technologies, train the workforce in skills needed for these technologies, ensure the ethical use of frontier technologies, collaborate with academia to drive research, and create a talent pipeline. The Academia should improve the curriculum by incorporating Frontier Technologies, and conduct research towards their applications. They should collaborate with the industry to reduce the skill gap, and explore interdisciplinary learning to understand the intersection of technology with other fields like economics, sociology, and law.

To conclude, I would say that given India's ambition of becoming a global technology leader, there's a strong impetus to develop emerging technologies from drug discovery and precision farming to mega factories for batteries and EVs. The industry of today increasingly requires higher degrees of automation. We need to accept the inevitable transition to a technology-led society and economy. More needs to be done to ensure that the benefits of technology are inclusive and the transition to a digital economy does not exacerbate inequalities. This was also discussed, so we should focus on skill development and reskilling, promoting innovation and startups, providing social security for displaced workers, building inclusive digital infrastructure, regulating new technologies, encouraging public-private partnerships, and fostering international cooperation.

I would like to take a moment to also dwell on the linkages between technologies and policies to drive growth. First and foremost, as policymakers, we must anticipatively project the transition that the labour force will experience, and accordingly roll out social security and re-skilling measures to make that transition smooth. Second, we need to reimagine workspaces in the era of automation and AI, creating new structures that draw out more human creativity and intelligence. Recall how the calculator first entered the office, how people thought it would affect them, and how much it really affected them. This brings me to the third link, pedagogy. There is no escaping the reality that our education system needs a massive overhaul, moving away from a mechanical and disengaged subject-based learning model to a dynamic and engaging systems-based learning model, with a strong emphasis on critical thinking.

The task is challenging, my friends, but with a proactive and forward-looking policy approach, the globe, as well as India, can leverage technology to create a future of jobs that is inclusive, sustainable, and prosperous.

References

1. Lutz W, Pachauri S (Eds.) (2023) Systems analysis for sustainable wellbeing. 50 years of IIASA Research, 40 Years After the Brundtland Commission, Contributing to the Post-2030 Global Agenda. International Institute for Applied Systems Analysis (IIASA)
2. IIASA Policy Briefs (2016) Analyzing population aging from a new perspective. IIASA Policy Briefs 12: https://iiasa.ac.at
3. Lutz W, Reiter C, Özdemir C, Yildiz D, Guimaraes R, Goujon A (2020) Skills-adjusted human capital shows rising global gap. Proc Nat Acad Sci United States of America (PNAS). https://doi.org/10.1073/pnas.2015826118[pure.iiasa.ac.at/17034]
4. Marois G, Gietel-Basten S, Lutz W (2021) China's low fertility may not hinder future prosperity. Proc Nat Acad Sci (PNAS). https://doi.org/10.1073/pnas.2108900118
5. Hoffmann R, Dimitrova A, Muttarak R, Crespo Cuaresma J, Peisker J (2020) A meta-analysis of country-level studies on environmental change and migration. Nat Clim Chang. https://doi.org/10.1038/s41558-020-0898-6
6. Rammelt CF, Gupta J, Liverman D, Scholtens J, Ciobanu D, Abrams JF, Bai X, Gifford L et al (2022) Impacts of meeting minimum access on critical earth systems amidst the Great Inequality. Nature Sust. https://doi.org/10.1038/s41893-022-00995-5
7. Piketty T (2014) Capital in the 21st century. Harvard University Press, London
8. Lutz W, Striessnig E, Dimitrova A, Ghislandi W, Lijadi A, Reiter C, Spitzer S, Yildiz D (2021) Years of good Life (YoGL) is a wellbeing indicator designed to serve research on sustainability. Proc Nat Acad Sci (PNAS). https://doi.org/10.1073/pnas.1907351
9. UNCTAD (2022) Friend-shoring and increasing concentration for global trade. https://www.hinrichfoundation.com/research/how-to-use-it/unctad-global-trade-update/
10. Laxenburg Austria. https://doi.org/10.5281/zenodo.8214208
11. World Economic Forum (2023) The Future of Jobs Report. https://www.weforum.org/reports/the-future-of-jobs-report-2023/

Chapter 3
Growth Implications of a Fractured Trading System

Peter Drysdale, Alicia Garcia-Herrero, Nagesh Kumar, Otaviano Canuto, and B. V. R. Subrahmanyam

3.1 Session Chair: Peter Drysdale

So, this session really is about what's happened to global trade in the last couple of decades, of course, really since the global financial crisis, trade dependency has leveled off, peaking at around 60% in 2008, 2009.

After the global financial crisis, it has fallen and hovered around 50 to 55%, which is a reflection of the falling relative shares of world trade and global output. And what the origins of this are rather important to making judgements about policy strategies down the track. And of course, this context informs a lot of the thinking about where to go next on a whole range of issues including climate change.

But development strategy as well. We've seen a significant increase in trade restrictions over that period, the last decade. The Global Trade alert records restrictive measures to have increased from about 9,000 a year to 35,000. And there's actually the measures that the IMF and others have set out under report the growing restrictiveness of international trade. A lot of informal measures and formal measures, especially anti-dumping acts, have become major instruments for trade protectionism.

P. Drysdale (✉)
Emeritus Professor of Economics and Head of the East Asian Bureau of Economic Research, Australian National University, Canberra, Australia

A. Garcia-Herrero
Chief Economist, Asia Pacific Natixis, Madrid, Spain

N. Kumar
Director and Chief Executive, Institute for Studies in Industrial Development (ISID), New Delhi, India

O. Canuto
Senior Fellow, Policy Center for the New South, Manaus, Brazil

B. V. R. Subrahmanyam
CEO, NITI Aayog, Delhi, India

S. Bery et al. (eds.), *Navigating Challenges for Sustainable Growth*,
https://doi.org/10.1007/978-981-97-7894-2_3

And then we have more recent developments towards industrial policy, the IRA in the United States, the Chips and Science Act, and developments in Europe that have also led to a tightening of the environment for an open international trading system. Let me make a few points about the sequence of developments that I think has shaped this environment over the last couple of decades in the United States.

But I don't think we've avoided some of the consequences of the significant recession that took place in the global economy around that period. And one of the consequences quite clearly was this increase in international protectionism, under the table, as it were to some extent, but increasingly explicit, and embodying formal measures that contravene the rules of the WTO and international trade. And an important impetus to that of course, was around the mobilisation of the politics of populist protection in Trumpian politics. And through that, the instigation of the China-US trade war, in which both US and China directly fought international trade wars, to the cost of competitive traders like Brazil and my own country.

And then, we've had on top of that, increased activism by China, aggressiveness by China in the imposition of trade coercion policies, for political purposes, including prominently in Australia. So the environment has degenerated significantly over that period, and these tensions in the international geopolitical sphere have fed into the trade sphere and led to the weaponisation of trade and economic policy more broadly, in an age of what can be called security driven industrial policy, especially in the industrial countries.

And broadly speaking, these are the causes of this downward shift in gear in global trade. The important thing to ask here, I think, is about the consequences and costs of this development.

How important is this development? Well, the consequences is that there's been an undermining of the multilateral rules-based system, alongside the long-term structural pressures on that, because the system hasn't encompassed fully a whole range of new areas that are important to the conduct of international trade and commerce.

And then with decoupling strategies, against China in particular as well as the policy measures taken around those strategies, we have witnessed increasing fragmentation of the system and a slowing of global trade and industrial transformation in the process potentially.

3.2 Speaker 1: Alicia Garcia Herrero

I'm here today to put together three difficult words: globalization, de-risking, decarbonization. Starting with a brief review of the de-globalization, I will focus on supply chains given my assumption that we will all reduce emissions. So I'm going to ask the question of how we decarbonise rather than whether we will. Therefore, supply, rather than demand, will be more of the focus.

If we look at the world global value chains (GVC) participation, signs of de-globalization have become increasingly clear since 2008 and it's a strong trend. In this de-globalization mode, the country that suffers the least is China, as its GVC participation remained relatively resilient. Europe, the world's most globalized region with the largest GVC participation, was highly integrated not only regionally, but also with the rest of the world.

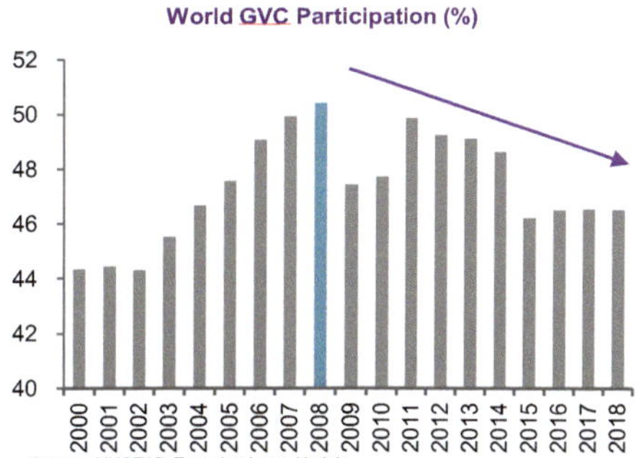

Source: UNCTAD-Eora database, Natixis
N.B. Results for 2016-2018 are forecasted by UNCTAD-Eora

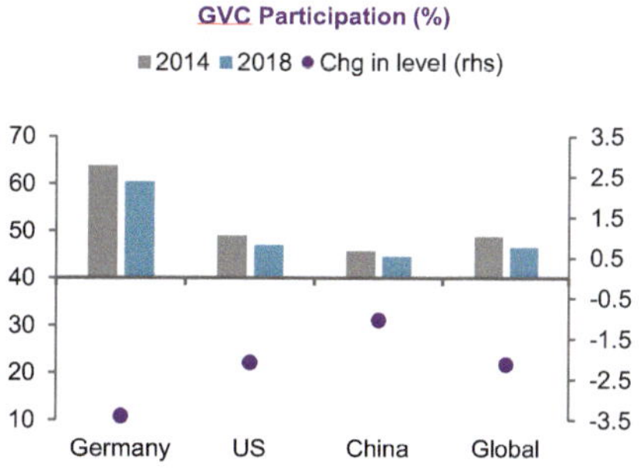

Source: UNCTAD-Eora database, Natixis
N.B. Results for 2016-2018 are forecasted by UNCTAD-Eora

The change in Europe's GVC participation was not obvious. But the fact is, Europe is losing a lot of its integration in the supply chains. This might be surprising even for the Europeans. In contrast, China is losing the least in supply chains integration, and the reason for this is twofold. First, China has continued to gain market share in global exports, acquiring 15% in 2021 while only around 1% in the 1980s. But another important reason is that China is increasing value added. Meaning, China is importing fewer intermediate goods to re-export and is exporting more of its value added for other countries to re-export. This is happening in China mainly in the green sectors, including solar, panels, EV batteries, and wind, where China manages to retain the value added. Therefore, China is more shielded than the rest of the world from the loss in globalization, especially regarding supply chains.

China has been dominating the battery manufacturing over the years. For solar panels, it is even more significantly concentrated on China. This concentrated globalization in the supply chain, which affects the world asymmetrically—more in Europe followed by the US, and much less so in China—is partly a factor behind the trade war. It does not mean it is justified, but it did reflect the asymmetric developments across the world in retaining value added. To mitigate the unbalanced development and excessive concentration, legislators and politicians have been rolling out policies in an effort to maintain their place in the supply chain. It is important for them to do so because this means jobs, innovation, as well as other important things.

Many countries have taken actions. These include not only the US, but also Europe with its emphasis on corporate sustainability due diligence, which seemingly related to human rights but encompasses much more. With Japan and South Korea also involved, this is apparently a widespread issue rather than merely US-centric.

A significant factor behind globalization, especially in the green tech sector, is technology-driven. This brings us to the US's push for decoupling, particularly in semiconductors. But the reality extends beyond that. US approval rate for export licences was much higher for the world than for China. The vast disparity there indicates not just fragmentation, but also a clear trend towards technological decoupling.

Moreover, the de-globalization trend is not limited to trade or supply chains. Global FDI flows, whether inward or outward, are also declining. And the trend is even more pronounced between the US and China, where much of the FDI was intended to support the supply chain. Evidence of de-globalization can also be found in finance. While China does hold a significant amount of US treasuries, such investments have decelerated rapidly in recent years.

The crux of my argument is the link between globalization and the primary concern in the world: the green tech, the sector where the supply chain is highly concentrated globally. It includes solar panels, wind turbines (though to a lesser extent), EV batteries, and encompasses different aspects from extraction, refining, intellectual property, to manufacturing.

In terms of extraction, China's concentration concerns far more than just rare earth metals. There are other metals where the concentration is even higher, and this does not even include China's ownership of mines abroad. Its dominance in extraction just came naturally due to its rich endowment in these resources.

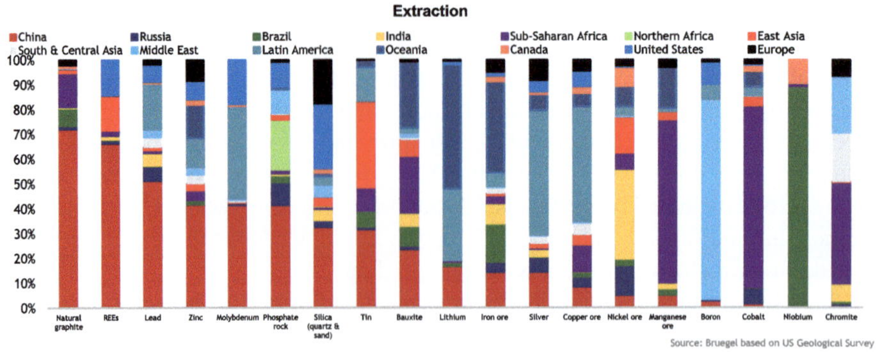

On refining, the concentration is even bigger, and forecast for 2025 shows a dominance around 90% across the board. On Innovation, China is also catching up quickly. Back in 2010, Europe was still leading the scientific development in most areas of the green tech, only except wind turbines, where China already gained its place. But moving into today, China is now dominating all innovation fronts, including solar, wind, batteries, heat pumps, and carbon capture and storage. For manufacturing, the concentration is even more apparent. China captures 90% of solar PV manufacturing, 60% of batteries, and 43% of wind turbines. It's a huge concentration on the manufacturing side as well.

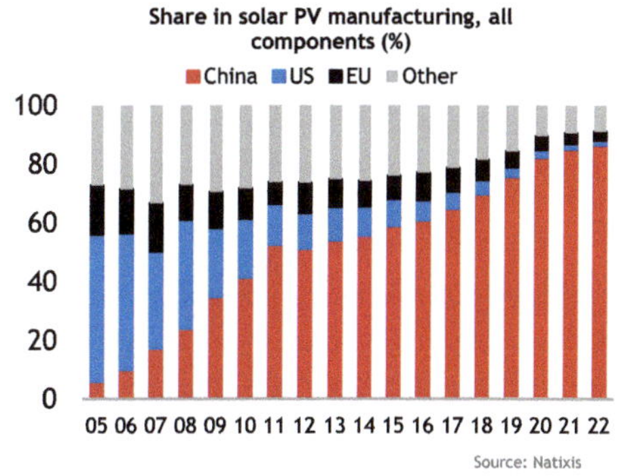

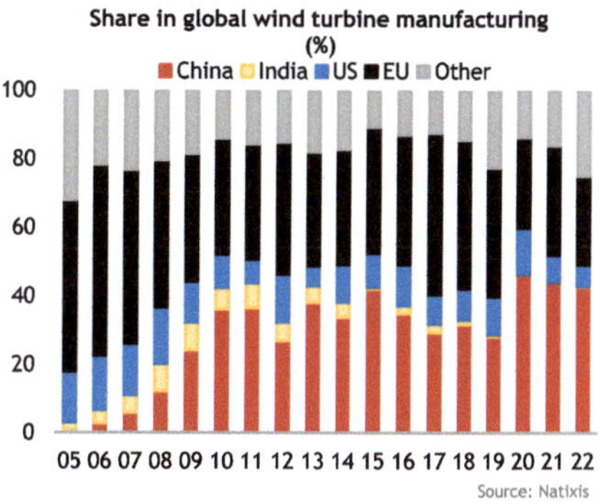

We are going to publish a policy note on this at Bruegel. We think that, with the assumption of China reaching net zero by 2060, it cannot produce enough for its own needs. But this does not correlate with China's manufacturing capacity, as it is already producing a lot. Rather, it is because of the huge demand that outpaced the supply. Meanwhile, we are relying on a single, or close to single, producer for the energy transition. This is somewhat irresponsible, from the rest of the world's perspective, as China also has a huge need domestically. The tremendous increase in demand, thus, appears risky, for what is becoming a monopoly on the import of inputs, such as critical raw materials, or the monopoly power on the export side.

These are unintended risks, which can be of various kinds, including China's growing domestic demand. And it is beyond the intended risk of retaliation from China that everyone, especially in Europe, seems to focus on. It is evident that the way we manage the green tech supply chains might be efficient, but not quite safe. Therefore, we are proposing a decarbonization partnership as an alternative solution. We're advocating this approach, aiming to present it to the EU council and to other global stakeholders including India.

The essence of the proposal is straightforward, and it extends beyond the G7's scope for the reason that the G7 alone cannot address the issues including the economies of scale required to establish a new supply chain. Our proposal is to create an additional supply chain, emphasizing coordinated specialization, which means leveraging the strengths of different nations. For instance, if India offers lower production costs for certain green tech products, it should focus on those. Our research indicates that countries such as India and some in Southeast Asia are highly competitive, nearly on par with China, and they can join certain parts of the supply chain where they enjoy comparative advantage.

The challenge, however, lies in ensuring technology transfer. That's where the concept of coordinated facilitation becomes essential. By having access to the required technology, regions with the capacity to produce can do so within a coordinated framework.

But given China's concentration and its cheap products, the question is now about how to incentivize countries to stop importing from China, and instead, turn to the additional supply chain that we are advocating? This is where cross-border carbon pricing comes to the forefront. Without doing so, it will never work because of lack of incentives. Therefore, aligning incentives is essential in this framework, in which there is an opportunity to break the monopoly on the inputs of critical raw materials as well as the near-monopoly on the export of green tech.

Will China lose? In our opinion, China wins out of this in any event, as it can at least preserve its decarbonization objectives.

In conclusion, we think slowbalization is happening in many areas, including trade, investment, and technology. Signs of fragmentation are showing in the supply chains and the reason why China was shielded in this trend is related to its dominance in the green tech sector. The current supply chain of green tech is neither enough nor safe for the world's decarbonization. An ideal one would be multilateral, in which multiple supply chains sprouting everywhere to mitigate concentration risks, but we have come to realize that it is no longer feasible as the world has seen the concentration of patents and China's low production costs. Therefore, policy action is needed to address the issue. But the US's IRA, Europe's Critical Raw Material Act or Net Zero Industrial Act won't work as they are too expensive. That is why we proposed an alternative solution, a partnership framework through incentive-aligned countries and using coordinated specialization.

3.3 Speaker 2: Nagesh Kumar

India has emerged as the fastest-growing large economy, in the post-pandemic era, as many leading economies of the world are facing a slowdown combined with persisting inflationary pressures while many others are reeling under the debt crisis. However, despite the slowdown, China remains the prime growth pole of the world economy, contributing 35% of global growth in 2023, with India contributing 15%. Can India emerge as the next growth pole of the world economy, leveraging its demographic and geopolitical sweet spots, as China's growth slows down due to its transition into an ageing society? This would be possible by sustaining an accelerated growth momentum. Realizing the Prime Minister's Vision 2047 for India to become a developed economy also would require sustaining a robust growth momentum for the next two and a half decades. Indian economy needs to grow at around 8% per annum for the next 25 years to realize this aspiration from the current 6–7% per annum. However, sustaining an accelerated growth rate becomes challenging with the external context turning less benign with a rather flat growth of world trade and the rise of protectionism that has turned globalization into 'slowbalization.' The question that this article tries to answer is what are the key opportunities, prospects, and policy priorities for sustaining India's growth momentum in a fractured trading system?

Growth Implications of the Fractured Trading System for India

Firstly, it needs to be recognized that globalization has been a mixed legacy and has had asymmetric gains for different countries. While China increased its share

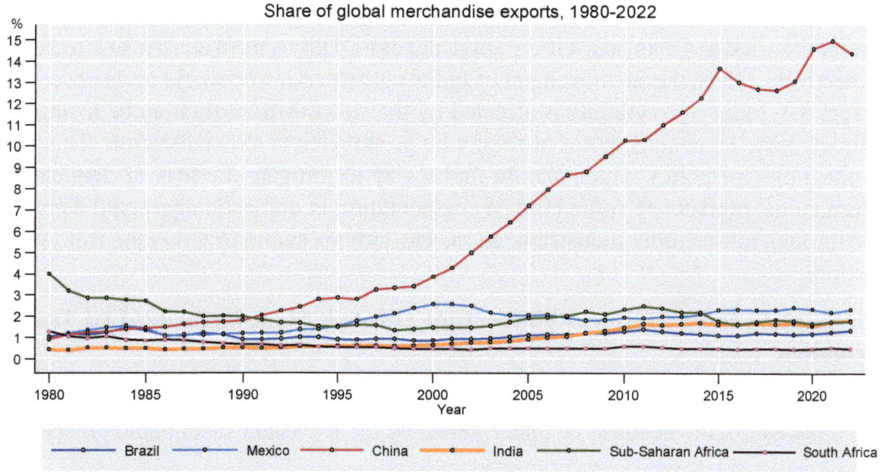

Fig. 3.1 Share of global merchandise exports, India, China, and selected developing countries and region, 1980–2022. *Source* ISID calculations based on World Development Indicators, World Bank, https://data.worldbank.org

in global exports from 1.79 to 14.36% between 1990 and 2022, other regions of the Global South had much more modest gains: India increased its share from 0.52% to 1.81%; Latin America and the Caribbean from 4.48 to 6.06%, while Sub-Saharan Africa was squeezed out with its share declining from 1.99% to 1.78% over the same period (Fig. 3.1). China was able to exploit the opportunities presented by hyper-globalization and capture a greater share of rapidly expanding global trade at the cost of others by quickly enhancing its manufacturing capacity.

The huge expansion of China's manufacturing capacity was a result of heavy strategic interventions. As documented extensively in the literature, the Chinese manufacturing prowess was underpinned by undervalued exchange rates, direct subsidies, local content regulations, among other strategic interventions.[1] Furthermore, China has been sustaining growing trade and current account surpluses over the years, sucking the global demand did not help other countries expand exports of manufactured goods to its large and growing market. In contrast, India has been sustaining growing merchandize trade deficits over the years, providing markets to other countries. Hence, the rise of India can be seen as a global public good. In that context, the ongoing decoupling and restructuring of the supply chains of global corporations on China +1 basis, presents an opportunity for India and other countries in the Global South to expand their global footprints.

Even though India hasn't integrated deeply with global value chains or benefited significantly from globalization, slowbalization is bad news for India's economic growth. The slowdown of global trade and investments (as summarized in Fig. 3.2 and Table 3.1) since the Global Financial Crisis of 2008/09 is very dramatic and sharp

[1] [1, 2].

with average annual growth rates of world trade coming down from 16.41% in the pre-GFC period to 4% in post-GFC period and of FDI flows from nearly 20% to 2.2% respectively. Given the co-movement of India's growth rate and world trade observed in Fig. 3.3, India's growth rate is affected by the slowdown. India may be losing an estimated one percentage point of economic growth due to slowbalization. This has implications for policy which should find a way to mitigate the loss of demand in international markets by some kind of augmentation in domestic aggregate demand and through job-creating industrialization. The latter is summarized in the following section.

Accelerating India's Growth Momentum Through Manufacturing-Led Transformation

The realization of Vision 2047 of developed country status and a US$ 5 trillion economy by 2026/7 needs to be underpinned by inclusive and sustainable prosperity

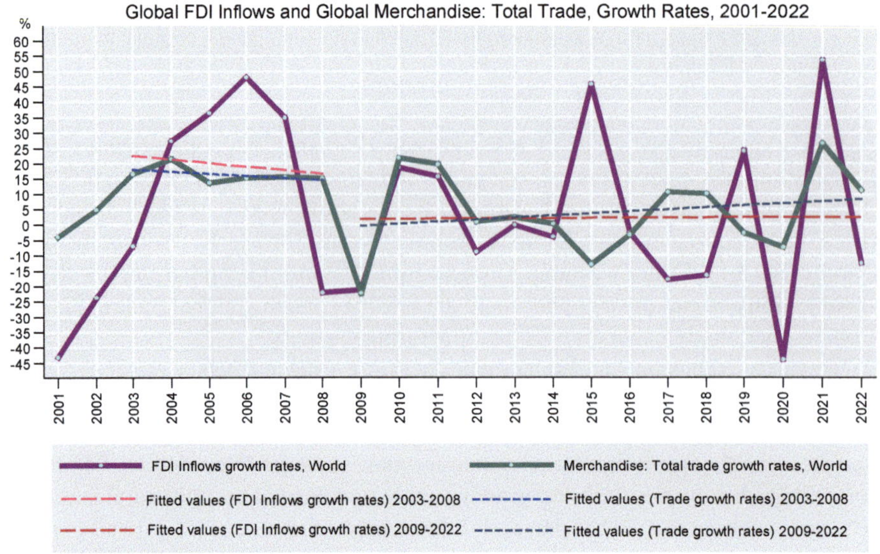

Fig. 3.2 Global FDI inflows and global merchandise: total trade, growth rates, 2001–2022. *Source* ISID based on UNCTAD STAT and *World Investment Report*, 2023, https://unctadstat.unctad.org/data, https://unctad.org/system/files/official-document/wir2023_en.pdf

Table 3.1 Average growth rates of global FDI inflows and Global Merchandise Trade, 2003–2022

	2003–2008	2009–2022
Global FDI inflows	19.65	2.22
Global merchandise: total trade	16.40	4.00

Source ISID based on UNCTAD STAT and *World Investment Report*, 2023, https://unctadstat.unctad.org/data, https://unctad.org/system/files/official-document/wir2023_en.pdf

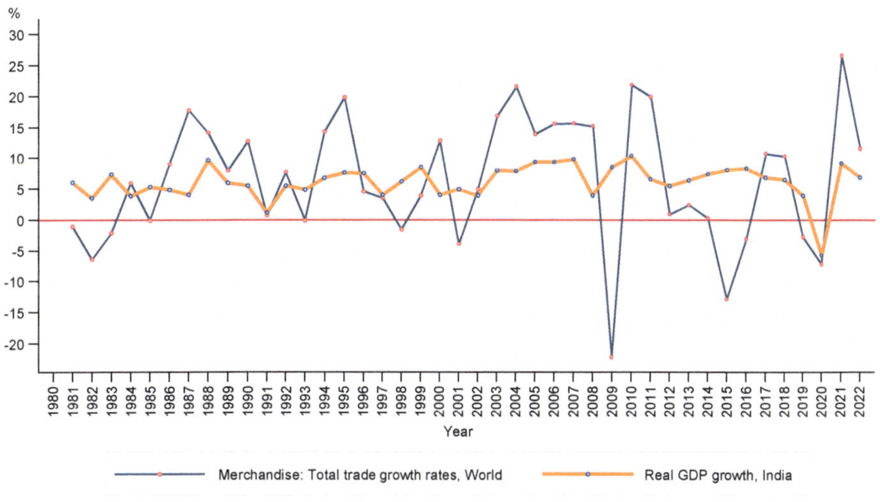

Fig. 3.3 Growth rates of World Merchandize Trade and India's GDP, 1980–2022. *Source* ISID based on UNCTAD STAT and the World Development Indicators

for all citizens through the creation of decent job opportunities for India's youthful workforce. The inability to create an adequate number of decent jobs in the past has led to nearly 86% of India's workforce getting locked in the informal sector without adequate social protection and remaining vulnerable to any shocks. The issue of decent job creation is linked with structural transformation associated with economist Arthur Lewis, where workers move over time from low-productivity activities (such as agriculture) to higher-productivity sectors (such as industry and services). India has witnessed the transformation of an agricultural-dominated economy into a services-dominated one bypassing the industry. While the service sector has delivered robust growth rates, it has not been able to absorb workers especially the unskilled and semi-skilled ones, in a proportionate manner. As a result, agriculture continues to sustain as much as 46% of India's workforce with barely a 15% share of GDP (Fig. 3.4). This services-oriented structural transformation, as it has been termed, has been able to absorb only 26% of the workforce in services. The manufacturing sector has been bypassed with its share in GDP stagnating at around 16–17% in contrast to an average of 30% in the East Asian countries. Not only has the share of manufacturing stagnated in India, but there is also evidence of some deindustrialization taking place (see [3–5]). The neglect of manufacturing to underpin the structural transformation in India has cost the country dearly in terms of creating decent jobs. The manufacturing sector has the highest backward and forward linkages compared to any other productive sector (Fig. 3.5). Hence, it generates more jobs indirectly for every direct job created.

It is for this reason development states across the world promote manufacturing sector. History corroborates that few countries if at all have attained prosperity without industrialization [6] has also argued persuasively that the growth of manufacturing not only drives economic growth but also enhances the productivity of the

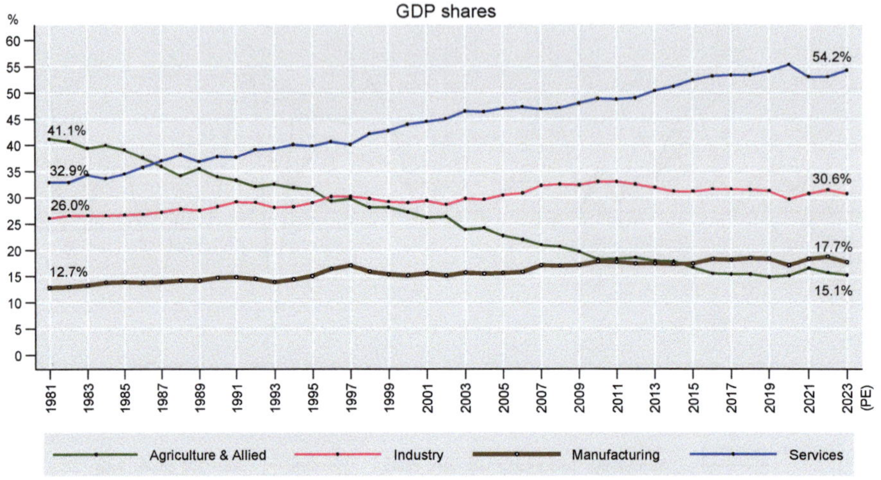

Fig. 3.4 GDP share, by sector, India, 1981–2023 *Notes*: 1981: FY1980-1981. 2022–23: Provisional Estimates (PE). *Source* ISID calculations based on National Accounts Statistics

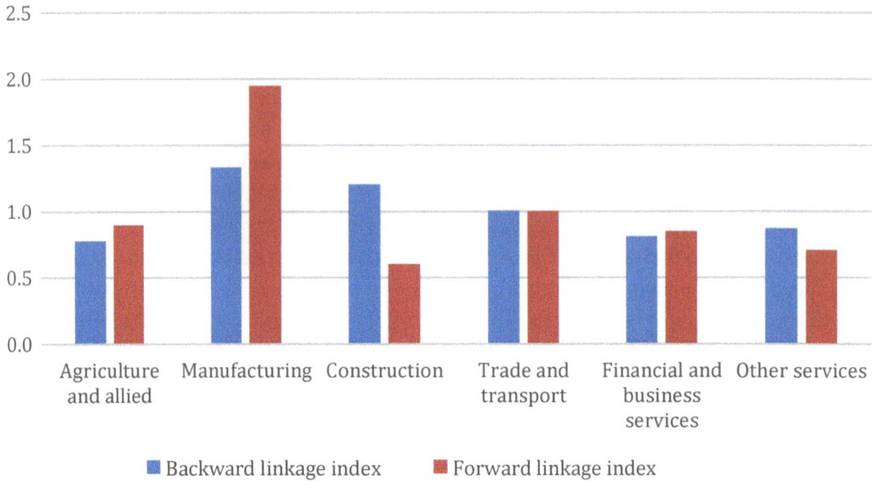

Fig. 3.5 Backward and forward linkages generated by economic sectors in Indian economy. *Source* ISID computations based on India's Input-Output Tables

economy overall with increasing returns to scale which could be dynamic in nature. The Agenda 2030 on Sustainable Development adopted at the United Nations Summit in September 2015 comprising 17 Sustainable Development Goals also recognizes the transformative potential of the industrial sector and seeks to enhance its share in employment and GDP (SDG-9.2). By substituting imports or expanding exports, an expanded manufacturing sector could also help to make India's balance of payments (BoP) more sustainable—which tends to periodically get into stress.

Therefore, faster job-creating rapid economic growth through manufacturing-oriented structural transformation, complementing the robust growth of the services sector, is the key to inclusive and sustainable prosperity of India for the realization of its Vision 2047 of a developed country. In that context, the Make-in-India programme announced by Prime Minister Modi in 2014 which seeks to tap the potential of manufacturing for India's development, was timely. It was further reinforced by *Atmanirbhar Bharat Abhiyan* in 2020 as a strategy to pull the economy out of the Covid-19 pandemic comprising a production-linked incentives (PLI) scheme to boost local production in 14 sectors.

The 'New Washington Consensus' on Industrial Policy

In achieving a manufacturing-led economic transformation, India could learn from the experiences of the industrialized and East Asian countries in fostering competitive manufacturing capacities through extensive state interventions. The developmental role of the State in these countries and the aspects of strategic interventions deployed that are collectively called industrial policy have been well documented in the literature (see [7], for a review). After becoming a bad word in the heydays of globalization, industrial policy is back in fashion across the world. Among many trends that the slowbalization and the Covid-pandemic have accentuated is a shift towards a real economy comprising production, jobs, and localization replacing the earlier emphasis on finance, consumerism, and globalization [8] has termed this trend 'Productivism Paradigm.' Governments around the world are adopting the so-called industrial policies that incentivize domestic manufacturing to create jobs and reshoring of value chains. The New Washington Consensus is not about liberalization and free markets. It is about industrial policy. A widely circulated IMF paper[2] *The Return of the Policy that shall not be named: Principles of Industrial Policy,* issued in 2019, recognized the 'strong commonalities in policies pursued by the Asian Miracles, and one cannot ignore the preeminent role of industrial policy in their development.' Over the past few years, there has been a deluge of evidence and debates on the relevance of industrial policy tools employed with varying degrees of success by traditional and late industrializers [9]. An extensive new review of evidence and experiences has concluded that 'there is a generic and powerful economic case for industrial policy and that the usual critiques rely on practical rather principled objections' and that the debate on industrial policy should be focused not on 'the whether' but on 'the how' [10].

The aggressive manner of adoption in recent times of industrial policy by some of the most advanced economies is a case in point. For instance, in the US, once the greatest champion of free markets and globalization, the Biden Administration has defined its industrial policy recently with the $280 billion CHIPS and Science Act, the $737 billion Inflation Reduction Act, and the $550 billion Infrastructure

[2] Cherif and Hasanov (2019).

Investment and Jobs Act. These Acts will foster local manufacturing and innovation of semiconductors chips, electric mobility, and other new technology products through hundreds of billions of dollars in subsidies and tax breaks. The European Union has followed suit with its own set of incentives and support for local producers. The new 'Green Deal industrial plan for the net-zero age' of February 1, 2023, sets out a European approach to boost the EU's net-zero industry, through measures to improve the competitiveness of the EU's net-zero industry including the 'net-zero industry act' of 16 March 2023, which aims to simplify the regulatory framework for production of key technologies, set targets for EU industrial capacity in 2030. One major outcome of the EU's climate-focused industrial policy includes the European Battery Alliance, a network to coordinate research and subsidize battery manufacturing across the continent [11]. EU is also looking to increase its share of the global semiconductor market and lead the way in quantum computing. Furthermore, the EU in December 2022 decided to impose a Carbon Border Adjustment Mechanism (CBAM), which will initially apply to imports of certain goods and selected precursors whose production is carbon intensive such as cement, iron, steel, aluminium, fertilizers, electricity, and hydrogen. EU importers must pay for emissions by buying CBAM certificates. The policy is set to take effect in 2026, with a transitional phase starting October 1, 2023. The policy is widely seen as unilateral, protectionist, and discriminatory adopted to safeguard domestic businesses [12].

India's Sweet Spots for Manufacturing

Disruptions in supply chains such as those following the COVID-19 pandemic and the Ukraine War have pushed global corporations to gradually de-risk their supply chains by diversifying them on China +1 basis. The restructured production is being directed it to friendly countries, termed friend-shoring. The IR4.0 is also a possible driving factor. In the past, global value chains (GVCs) were outsourced to developing countries to leverage labour cost differences among other locational factors [13]. Robotization of production driven by IR4.0 tends to neutralize the labour cost advantage enjoyed by developing countries. The reshoring of global value chains is, therefore, a real possibility and can affect the export prospects of developing countries [14].

One could argue that India's recent manufacturing push through various industrial policy instruments is a part of the global trend of governments incentivizing domestic manufacturing to create jobs and re-shore value chains. India will be helped by its position as a 'geopolitical sweet spot,' having friendly relations with key industrial countries in the West and East. This will allow India to benefit from global companies' friend-shoring supply chains to diversify them away from China.

India is also enjoying a 'demographic sweet spot' with a relatively young population. The proportion of the working-age population in India will peak at 68.9% around 2030 and will stay favourable for a few decades. This contrasts rapidly ageing populations in most industrialized countries such as Japan and European countries as well as newly industrialized countries such as the Republic of Korea and China (Fig. 3.6). The youthful population also makes it possible for the country to train

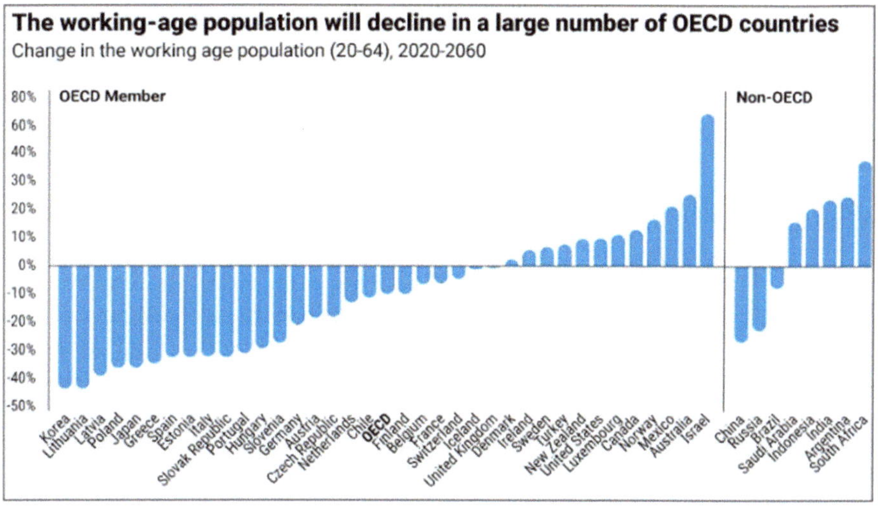

Fig. 3.6 Changing proportions of working age population, 2020–2060. *Source* World Economic Forum, https://www.weforum.org/agenda/2020/02/ageing-global-population

them in emerging disciplines such as machine learning among other artificial intelligence (AI) tools to harness the emerging IR4.0 technologies for its development besides catering to the global skills requirements, becoming the talent capital for the world (see [14]).

Opportunities for Manufacturing-Led Transformation

As India strives to build competitive manufacturing capabilities, an important question would be: What opportunities are available to India in terms of feeding the domestic demand versus external markets and emerging opportunities? Given below are a few pointers for these opportunities.

1. Making for India

The biggest opportunity for expanding the country's manufacturing base is by producing for domestic consumption. One should start by reversing the trend of the rising share of imports in final consumption, as Indian companies outsourced production offshore to save costs in the decade following 2004 with an appreciation of the rupee.[3] Outsourcing has been practised widely by several well-known Indian companies by getting their products manufactured in other countries, mainly China, and then continuing to sell them under their brand names. Outsourcing of production was practised even for several price-sensitive home electrical and electronic appliances (electric fans, toasters, mixer-grinders, juicers, wall clocks, TVs, refrigerators, air-conditioners, etc.) that used to be manufactured in the country for many decades. Reversing this trend of hollowing-out of the Indian industry is the first step towards industrialization.

[3] [4] for evidence.

Then there are other industries with significant import dependence such as power equipment, electronics, a variety of organic and inorganic chemicals, and active pharmaceutical ingredients (APIs), that can be manufactured within the country as adequate domestic demand exists. The PLI schemes announced by the Government as a part of the *AatmaNirbhar Bharat* package in 2020 are trying to incentivize domestic production of some of these products. Considering that India's manufactured imports add up to $370 billion per annum (out of the total imports of around $750 billion in 2022–23), the substitution of even 50% of the manufactured imports in a gradual manner could enhance the current scale of manufacturing value added of roughly around $550 billion per annum by 33%. Therefore, there is considerable potential for strategic import substitution. Growing demand for consumer and capital goods and defence equipment would continue to provide additional opportunities for the local manufacturing base with scale economies. The competitive manufacturing plants exploiting scale economies would also be able to tap opportunities that may arise in the international markets.

2. Making for the World or Export-Oriented Manufacturing

Notwithstanding the slowbalization and rising protectionist trends, India is likely to benefit from the strategy of global corporations to de-risk their supply chains by diversifying them on a China +1 basis. This reshoring is likely to help India get integrated with the global and regional value chains. Furthermore, strengthening India's presence in traditional areas such as textiles and clothing, leather goods, gems and jewellery, processed foods, vaccines and generic pharmaceuticals, automobiles and components, refined petroleum products, steel and non-ferrous metals, and some types of machinery and electrical equipment is vital, besides making inroads in new areas and markets. Given India's rather marginal 1.7% share of global merchandise exports, even a very small rise of 0.5% in this share over the next 2–3 years will add US$ 100 billion to India's exports and possibly US$150 billion to manufacturing value added (MVA).

3. Sunrise Industries: Electronics and Semiconductors

The digital revolution also provides fruitful opportunities for fostering manufacturing in India. India can leverage its unique strengths such as its pool of technical manpower, software and chip design capability, and large domestic market to exploit these opportunities. Annual imports of electronics are of the order of $80 billion and are growing rapidly with projections of $400 billion of imports by 2025. Emergence of India as the net exporter of mobile handsets since 2022 is an important development with Apple and Samsung assembling their mobile handsets in India in an increasing manner. However, the value addition of the handsets assembled in the country needs to be enhanced. In that context, Chinese vendors of Apple allowed to establish joint ventures to produce components. The recent government initiatives to develop design, manufacture, and export semiconductor chips including through US$ 10 billion Semiconductor Mission to foster manufacture of semiconductor chips and displays leveraging India's leadership in software development and chip design. This has led to some credible investment proposals for semiconductor chips and displays, which, if successful, could transform the whole electronics ecosystem while reducing import dependence. The manufacture of semiconductors in the country will help to catalyse the electronics ecosystem comprising a whole range of downstream products.

4. Sunrise Industries: Green Industrialization

A whole new range of green industries primarily driven by India's ambitious targets of clean energy transition with 50% of energy coming from renewable sources by 2030. These targets are driving manufacture of green and blue hydrogen, solar panels, and wind turbines. Government is also promoting electric mobility and energy efficiency which is leading to rising emphasis on production of electric vehicles (EVs) and two wheelers. Electric mobility is creating a rising demand for Li-Ion batteries and other storage solutions. All of these sectors offer very promising industrialization avenues while also advancing the sustainability agenda. India should aim to become a global hub of compact EVs (including two and three-wheelers) and batteries. The Government has also announced a $2.3 billion Green Hydrogen Mission with objective to make India a leading manufacturer and exporter of green hydrogen. These new green industries will not only help to create jobs and incomes but also advance India's Net Zero target.

To sum up, translating these opportunities for strategic import substitution, export promotion, and digital and green industrialization has the potential to lift India's MVA from the current $550 billion to $1 trillion by 2026/7, thus advancing the government's $5 trillion economy target and creating millions of decent jobs in the process. Manufacturing value added could reach US$ 7.5 trillion out of the projected GDP of $30 trillion in 2047.

A Policy Agenda for Manufacturing-Led Transformation

What could be the policy lessons for accelerating growth of the manufacturing sector for decent job creation, complementing robust services sector growth, as an additional engine to power India's transformation towards a developed economy and emergence as a global growth pole? Over the past decade, the government has taken several reforms to tap the potential of the manufacturing sector. Recognizing that manufacturing sector development requires a conducive policy ecosystem or industrial policy, covering promotional measures including incentives, a supportive trade and exchange rate regime, finance and credit, an innovation-friendly intellectual property regime, and supportive physical and social infrastructure, implemented in a coordinated manner, as summarized below.

Beyond PLI: Pro-active Investment Promotion and Incentivisation of Manufacturing Investments: As part of Make-in-India, the Government has focused on improving the ease of doing business (EODB) in India through the abolition of obsolete regulations and processes that hindered industrial investments. The government has also increased FDI ownership limits in a number of sectors—such as railways, defence manufacturing, insurance, medical devices—and created an investment promotion and facilitation agency, *Invest India*. Import tariffs were raised in select sectors to give some infant industry protection. The corporate tax rates were lowered especially for new enterprises. Major reforms such as the Goods and Services Tax (GST) which make India as a single market for the first time and the Indian Bankruptcy Code (IBC) which provided a framework for resolution of non-performing assets of the banking sector were introduced. As a result of these steps, India's place in the World Bank's EODB rankings moved up sharply from 142 in 2014 to 63 in 2019 (before the index was abandoned in 2021). India has started to attract greater magnitudes of FDI inflows, which crossed a record figure of $81 billion in 2021–22 [15]. India has also developed the third largest ecosystem for Start-ups in the world with nearly 100,000 recognized start-ups of which more than 100 have become unicorns.

Make-in-India was reinforced in a big manner by the production-linked incentives (PLI) scheme introduced in 2020 as a part of the *Aatmanirbhar Bharat* package announced to revive the economy in the aftermath of the Covid-19 pandemic. The PLI scheme provides a 4–6% incentive to boost local production (or substitute imports) and exports for 14 select sectors. These include sunrise and green manufacturing products, such as solar photovoltaic cells and modules, advanced chemistry batteries, active pharmaceutical ingredients, large-scale electronics, medical devices, specialty steels, and telecom and networking equipment. In an effort to create a full ecosystem of electronics, the government launched in 2022 a $10 billion Semiconductor Mission to foster the manufacture of semiconductor chips and displays. Also in 2022, the government announced a $2.3 billion Green Hydrogen Mission with the objective to make India a leading manufacturer and exporter of green hydrogen.

Given that all the major governments of the world including the US and EU are offering investment incentives to attract investments, incentives of the type that are offered under PLI for a fixed term are desirable to build industrial capacities and help to scale them up. The early results have been encouraging. India has turned into a net exporter of mobile handsets after being a net importer. Monthly exports of India-assembled mobile handsets crossed $1 billion in September 2022. There are indications that Apple could be sourcing 25% of its handsets from India by 2025, up from under 5% at present. Leading Indian energy companies have also committed large investments in the manufacture of green hydrogen. There are also some credible proposals for the manufacture of semiconductor chips and display devices, including by Foxconn. However, PLI has not been able to attract investment in a number of other sectors, forcing the Government to review the scheme. It may also be desirable to link the incentives offered under PLI to progressive value-addition rather than just value of production or sales.

Furthermore, investment promotion, especially of FDI inflows should go beyond marketing and facilitation to proactive targeting which can help to attract investment inflows of better 'quality' than those that enter on their own. Proactive targeting requires the investment promotion agency to have a strong research and analysis department that will help it to identify areas where the size of domestic demand and/or the country's other advantages/resources justify localization of production in a competitive manner. It would go on to develop viable investment projects to entice potential global corporations to invest in them including through requests for proposals (RFPs). The RFPs would enable the country to obtain the best terms from rival MNCs in terms of deepening value addition, export promotion, vertical inter-firm linkages and vendor development, and transfer of technology among others. A case in point is the orders placed for 1200 civilian passenger airliners by Indian carriers in 2022–23 which exceed the combined annual commercial aircraft production of both Airbus and Boeing. India's investment promotion agency could have used this opportunity of having a relatively large domestic demand to invite both Airbus and Boeing to bid for an assembly line of single-aisle jet aircraft to be set up in India, offering them some facilities such as land and incentives but also some performance requirements. The assembly of aircraft in the country could unleash an ecosystem for several ancillary units that supply parts to the aircraft makers. The investment promotion agency could also examine how the offset conditions attached to the government procurements be best exploited by proposing the development of a vendor base, among other possibilities.

Manufacturing Sector Needs to Be Supported by a Specialized Financing Institution: Access to affordable credit is critical for industrial development. Hence, in European countries as well as in the East Asian countries the governments have intervened to ensure easy access to affordable credit to foster industrialization, as clear from experiences of Germany with KfW, Brazil with BNDES, South Korea with KDB, and China with CDB. In India too, a trinity of development financing institutions, namely, the Industrial Finance Corporation of India, the Industrial Development Bank of India, and the Industrial Credit and Investment Corporation of India played an important role in providing term-lending to the industry till 2001 when they were made to convert themselves into commercial banks as a part of financial sector reforms (See [16, 17]). Thus, since the turn of the century, industrial credit has been primarily catered to by commercial banks especially the public sector banks for both long- and short-term investment needs. The corporate bond market which plays an important role as a source of long-term finance to industry worldwide, has failed to develop in India despite decades of reforms and lacks depth and scale. After the initial spurt in the 1990s, the importance of the stock market as a source of capital also declined. Enterprise surveys do corroborate that finance for investment is a constraint faced by them, especially for small and medium-scale firms forcing an overwhelming proportion (70–75%) of them to rely on internal sources. Commercial Banks remain ill-equipped for term-lending due to asset-liability mismatches and lack of technical expertise.

As observed earlier, India's 2047 Vision requires it to grow around 8% p.a. which would require the manufacturing sector to grow at around 9–10% p.a. for the next 25 years. To catalyse staggering investments in the manufacturing sector needed for sustaining accelerated growth, ISID (2023) has proposed the creation of a new DFI for the industrial sector namely, the *National Industrial Development Bank of India* (NIDBI), besides strengthening the corporate bond market. The creation of NIDBI will help drive industrialization by addressing the gaps in the existing industrial financing system in tune with national priorities. It could also develop specialized expertise on project appraisal, risk management, impact assessment, and keep track of emerging developments globally and responses needed at the national level.

Address the Vulnerability of MSMEs: Contributing nearly a third of India's GDP, MSMEs have been an important engine of economic growth. However, MSMEs face several constraints and remain vulnerable to shocks given their small scale of operation, weak financial status, and unorganized nature. MSMEs have been affected by the liberalization of India's trade policy since 1991, particularly after the removal of quantitative restrictions in 2000–01. The sharp rise in the import penetration of a number of consumer goods including handicrafts among other labour-intensive goods generally produced by MSMEs may have affected their growth prospects. An ISID study found that consumer durables/non-durables predominantly accounted for the major surge in imports leading to the share of consumer goods imports in total imports of India rising from 11.7% in 1996–97 to almost 19% in 2019–20.[4] Organized retail and e-commerce companies have become an important conduit of imported goods in the consumer goods space as corroborated by the import figures of single-brand as well as multi-brand retailers. Although India has frequently deployed safeguards against import surge provided under the WTO Agreement on Safeguards and the WTO Agreement on Anti-Dumping, MSMEs may have been handicapped in seeking government intervention being unorganized and lacking strong lobbying ability. Besides promoting and adopting sector-specific policies such as those adopted for toys and games that have brought about desired results over the past couple of years, it is also important to harness the potential of the organized retail sector. The marketing power of organized retail and e-commerce players may be leveraged by imposing a performance requirement on them to develop their local vendor base among MSMEs for exports commensurate with their imports of consumer goods. This would also help Indian MSMEs integrate with global value chains.

Innovative Activity of Enterprises Needs to Be Strengthened: India has moved up the global innovation rankings from 81 in 2015 to 40 by 2022. Thanks to its leadership in ICT software, India's AI preparedness is considered to be relatively high. India has also emerged as an important base for R&D by MNEs hosting 1600 global capability centres (GCCs). However, at 0.7% of GDP (although possibly an underestimate), India's R&D expenditure is rather low compared to other emerging countries. Furthermore, only 39% of this expenditure is undertaken by the industry and the bulk is spent by the public-funded national research laboratories. Indian industry needs to scale up R&D activity sharply if it is to emerge as a significant player as a manufacturing hub and to leverage new technologies such as artificial intelligence and machine learning (AI/ML). In that context, the establishment of the National Research Foundation (NRF) is an important initiative, which could hopefully power the innovative activity of Indian enterprises. The government could also consider restoring weighted deductions for R&D expenditure by companies, especially for incremental R&D intensity. India can also consider adopting a second-tier patent system e.g. Utility Models or Petty Patents that provide limited protection for incremental innovations. Utility Models have been extensively used by East Asian countries to foster incremental innovations. Utility Models may particularly encourage innovative activity of MSMEs that is of generally incremental nature.

---

[4] Arun [18].

Closing Gaps in Industrial Infrastructure and Logistics: Indian Government is implementing an ambitious plan of logistics infrastructure and industrial corridors to obviate infrastructure constraints as well as to provide efficient logistics infrastructure to facilitate industrialization. This includes the National Industrial Corridor Programme covering include Multi Modal Transport Network—Railways, Highways, Expressways, Waterways, Airports, and Ports; Logistic/Transhipment Hubs; Industrial Cities/Townships and Urban Infrastructure sometimes termed as FIRE Corridors (Freight, Industrial, Railways, and Expressways). There are 11 industrial corridors underway covering the length and breadth of the country. The first one, Delhi Mumbai Industrial Corridor (DMIC) is the most advanced in terms of implementation. The programme is now a part of the $1.2 trillion National Master Plan for Multi-modal Connectivity launched in 2021. However, the corridor development has been slow having been affected due to many constraints such as land acquisition, the environment/forest clearances, legal disputes, delayed construction by some States, Covid-19 pandemic related lockdowns, and by poor coordination within and between States. Coastal Economic Zones and Ports-led development also has important potential. Around 95% of India's trade by volume and 70% by value passes through ports. Areas in and around ports are attractive industrial locations in view of their easier access to global markets. This explains higher concentration of investments and industrial agglomeration in the coastal regions. The Government of India under the Sagarmala Perspective Plan has identified 14 areas to be developed as Coastal Economic Zones. It has also identified 30 potential clusters to be developed in these zones including power generation, refineries and petrochemicals, cement, electronics, apparel, leather, furniture, and food processing. As many as 240 of the 377 SEZs are also located in coastal states. The ongoing initiatives have helped improve India's place in the World Bank's Logistics Performance Index by 6 places to 38 out of 139 countries in 2023.

Making India's Education System and Skill Development Fit-for-Purpose: The availability of skills is a critical ingredient for success in industry. India needs to completely revamp the educational system to produce the type of skills that are needed including for the incipient digital revolution. The government is also paying attention to skill development through the Skill India Mission. The National Skill Development Corporation is approaching the skill gaps by expanding public–private collaboration, initiating pathways for international mobility, and increasing women's participation in the labour force. Given the growing scarcity of skills that are fit-for-purpose for AI/ML, the industry is learning to reinvent strategies for recruitment, training, and retention of talent.[5] The government is also paying attention to skill development through the Skill India mission. The National Skill Development

[5] See for instance, https://www.livemint.com/technology/tech-news/indian-it-cos-struggle-to-fill-digital-skills-gap-11626025655102.html; https://economictimes.indiatimes.com/tech/information-tech/inside-the-war-for-tech-talent-in-india/articleshow/88677638.cms; https://economictimes.indiatimes.com/tech/information-tech/it-firms-battle-attrition-with-tech/articleshow/90742792.cms.

Corporation is approaching the skill gaps by expanding public–private collaboration, initiating pathways for international mobility, and increasing women's participation in the labour force. This would include revamping secondary and higher education to design thinking and problem-solving and introducing coding in schools, besides improving the quality of education at all levels. The seats in secondary schools, colleges, and higher education institutions need to be rebalanced in favour of Science, Technology, Engineering, and Mathematics (STEM) vis-à-vis traditional humanities and arts disciplines. Even within IITs and other engineering institutions, there is a need to rebalance the seats in favour of computer science, AI, data science, machine learning, and algorithm-related courses against traditional engineering disciplines such as civil, mechanical, or chemical engineering.

The National Education Policy (NEP) 2020 emphasizes multidisciplinary education, vocationalization, STEM, and strengthening technical education with a focus on cutting-edge areas like AI, big data analysis, and machine learning, among others that would be critical for harnessing IR40. It also envisages Digital Universities that would enable students to design a more personalized and flexible education. It also recognizes the need to avoid the commercialization of education and the importance of providing affordable quality education. These changes will help the Indian education system produce graduates who would be needed rather than those who cannot find a job. A big expansion in the public-funded education and training sector through raising the national education spending to the recommended 6% and above, from the current level of 4.4%, to provide affordable, quality education in the emerging AI/ML-related fields through a reformed education and skill development framework would pay rich dividends to the country in terms of harnessing the potential of IR4.0 for its own inclusive development. This could help the country become the skill capital of the world.

Exchange Rate Management: Exchange rate management has been an important industrial policy tool. East Asian countries have widely used managed exchange rates as a tool for fostering industrialization. Japan has extensively used the depreciated exchange rate of the yen to boost the competitiveness of its exports until the Plaza Accord of 1985. In the early years of industrialization, the Republic of Korea (ROK) rationed foreign exchange, giving priority to importers of capital goods and intermediate inputs [19]. The Chinese government adopted initially a dual-track exchange rate system, allowing the market-determined exchange rate to operate parallel with the overvalued official exchange rate, and the dual-track system converged to a managed floating system in 1994, followed by a hard peg during 1995–2005, allowing the exchange rate of yuan to move within a narrow band since 2005 [20]. The Indian rupee, on the other hand, has tended to appreciate in real terms over the years especially after 2004 despite the country consistently running current account deficits, due to significant short-term capital inflows. The rupee appreciation has led to the erosion of the competitiveness of Indian products thus encouraging the outsourcing of production by Indian companies for even their domestic markets

[4]. As a part of industrial policy, therefore, the RBI should be required to maintain a competitive and slightly depreciating exchange rate of rupee in real terms and vis-à-vis the major competitors in the export markets.

Augmenting Aggregate Demand Through Income Transfers: Normally industrial policy focuses on easing the supply-side constraints on industrialization. However, the demand side should not be overlooked. As earlier observed above, the slowdown of world trade and rising protectionism in the global economy, a phenomenon referred to as slowbalization has affected the growth rate of the Indian economy by an estimated one percentage point by reducing the demand for goods and services exported by India. The slowbalization is likely to be intensified with the threat of recession and stagflation looming large in the industrialized countries. In that context, some kind of augmentation of aggregate demand would be critical for sustaining robust growth rates of GDP in general and manufacturing in particular. To mitigate the loss of demand in the international markets, the Government may consider an income support scheme for the bottom 30% of the population. Besides stimulating economic growth and demand for manufactured products, such a scheme could also help to make a dent in the rising inequalities and persisting poverty, as recognized by the NITI [21] (also see [14]). The digital infrastructure to implement such a scheme is already in place. Income transfers under UBI can also be linked to some other social objectives, for example, by making it conditional to putting girls in school. The fiscal sustainability of this needs to be worked out but as it would do away with several social welfare schemes run by the Government, releasing resources that could be channelled into the new scheme.

Supportive Regional and Multilateral Trade Rules: India's opportunity to get integrated with the supply chains of global companies will be facilitated by its participation in a broader regional trading arrangement. Although India has free trade arrangements with key regional players such as Japan, Republic of Korea, the ASEAN, Australia, its inability to be a part of a regional agreement with cumulative rules of origin may affect its attractiveness for value chain integration adversely. From that point of view, India should consider joining either the Regional Comprehensive Economic Partnership of East Asia (RCEP), or the Comprehensive and Progressive Agreement for Trans-Pacific Partnership (CPTPP), or the incipient trade agreement under the Indo-Pacific Economic Framework (IPEF). Among the three options, the terms of joining RCEP may be less onerous compared to the other two because of their coverage of non-trade issues that may undermine India's development policy space. India has participated in the negotiations of RCEP throughout and may be able to negotiate a long transition or exceptions to protect its vulnerable sectors. Being part of the regional arrangement would certainly be an advantage.

Finally, in the context of G20 and international cooperation, there is a need to retrieve some of the policy space for pursuing industrialization and access to environmentally sound technologies. It may be helpful to have flexibility to deploy some performance requirements such as domestic content requirements (DCRs) which have been used extensively by the industrialized countries in the West as well as in the East but were withdrawn under the WTO's TRIMs Agreement. India's use of DCRs for solar PVs was successfully challenged in WTO in 2015–16. However, in 2022, President Biden of the US authorized the use of the Defence Production Act and super preferences under the Buy America Act including DCRs among other actions to spur local manufacturing of solar power equipment. Hence, India should develop a consensus in G20 and other forums regarding policy flexibility to deploy the use of DCRs for building its renewable energy equipment industry through a 'peace clause.' Similarly, there is a need to facilitate access to environmental technology by extending the patent waiver for such technologies following the precedence of TRIPs and Public Health adopted in 2003 which has helped to address the global AIDS challenge. India could also seek to define the provisions for the transfer of technology under Article 66.2 of the TRIPS Agreement, which has remained the best endeavour clause in the absence of detailed provisions.

Concluding Remarks

The foregoing analysis has shown that a manufacturing-led transformation is imperative for India to realize its development aspirations of building a developed economy by 2047 and to address the challenge of employment creation and sustainable management of balance of payments. Industrial policy and emphasis on the real economy is a global trend in the context of slowbalization, rising protectionism, and fractured trading system. As global companies restructure their supply chains on China +1 lines, India can potentially leverage its geopolitical and demographic sweet spots to build manufacturing capacities to feed growing domestic and global demand and tap the opportunities presented by the digital and green industrial revolutions. To tap these opportunities, a strategic approach is needed to harness the potential of manufacturing, for which many useful lessons are available from the experiences of East Asian countries. In that context, India should build on PLI to a more proactive targeting approach to investment promotion that would help to attract better quality investments meeting its development needs. The manufacturing-led transformation would also need to be supported by a specialized term-lending institution, by competitive management of exchange rates, efficient physical infrastructure, and logistics facilities. A fit-for-purpose educational and skill development system can not only feed the domestic requirements of skilled manpower but also has the potential to make the country a talent hub for the world in the context of ageing societies and the rise of Industry 4.0. The innovative activity of Indian enterprises has to rise sharply to enhance their competitiveness in international and domestic markets. MSMEs need to be integrated with the value chains of organized retail through performance requirements. Some augmentation of aggregate demand through conditional income transfers to the bottom 30% of the population could also help in addressing the rising income inequalities in the country. India also needs to make the regional and global trade rules supportive of its ambitious manufacturing-led transformation. Finally, to be effective, the different elements of industrial policy as outlined above need to be pursued in a coordinated manner. This would require a high-powered institutional architecture for a coordinated implementation of industrial policy.

3.4 Speaker 3: Otaviano Canuto

Dear colleagues.

What I will try to do is basically to illustrate something that Professor Peter has so aptly spoken. Namely, if we want to understand the growth implications, particularly costs of moving towards a fractured trading system, we must use as a benchmark exactly what happened over the period which is usually called as hyper-globalization or globalization 2.0.

So, what I'll try to highlight is exactly what aspects were those, so as to use them as a benchmark to think about the costs of a fragmentation of the trading system. What was hyper-globalization or, like Professor Richard Baldwin calls it, Globalization 2.0? In the eighties and nineties, particularly the nineties, we saw the manifestation of the outcome of what one may call a tectonic plate shift under the global economy. This was the combination of three things:

First, a cluster of technological innovations not only in information and communication technology but also in transportation. The containerization allowed the fragmentation of manufacturing processes to levels previously unthinkable.

Second, the reasonably widespread adoption of trade opening policies. Literally, in most countries in the world, particularly developing countries, there was a move towards reducing tariffs and non-tariff barriers to trade.

Third, the incorporation, almost overnight, of 1 billion workers with lower wage aspirations into the global supply of labour available for market economies. I'm referring here not only to the downfall of the Berlin Wall but also to President Deng Xiaoping's creation of special economic zones that allowed for a tremendous rise in exports and imports as a share of China's GDP.

The results? Well, there was substantial growth of GDP per capita in emerging markets and developing economies. The correlation between trade insertion in exports and increases in GDP per capita can be seen in Chart 1. And, as Professor Peter mentioned, there was a change in the composition of the global economy and trade, with the rising weights of not only China but also other emerging markets and developing economies.

This resulted in a significant reduction in poverty rates. At the same time, there was a double movement with respect to inequality. The world became less unequal when it came to per capita income, but there was a simultaneous rise in within-country inequality, particularly in advanced economies, as depicted in Chart 2. These were direct results of trade integration.

Along with higher foreign trade came the transfer and local absorption of knowledge and technology in machines, equipment, and also intangible forms, accompanying the formation of global value chains. This is evident, for instance, in estimates made by the IMF on how foreign knowledge contributed to labour productivity growth among advanced economies and also in emerging market economies. As shown in Chart 3, the IMF estimates that from 2004 to 2014, foreign knowledge accounted for about 0.7 percentage point of labour productivity growth a year, what

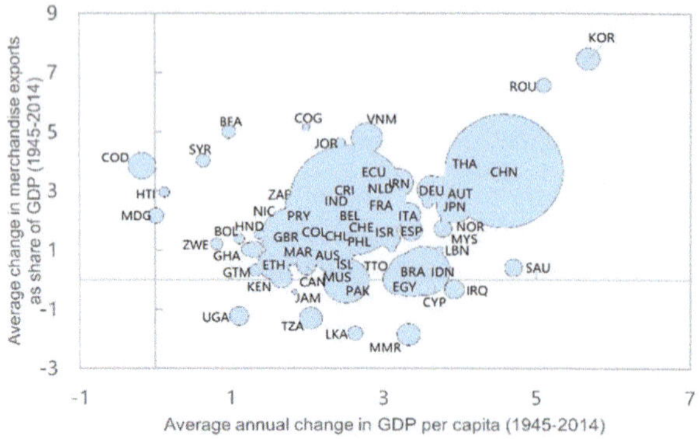

Chart 1 Growth of GDP and trade, 1995–2014. Average change in real GDP per capita versus average annual change in exports as % of GDP. *Source* Aiyar, S. et al. (2023). Geoeconomic fragmentation and the future of multilateralism, IMF staff discussion notes SDN/2023/001

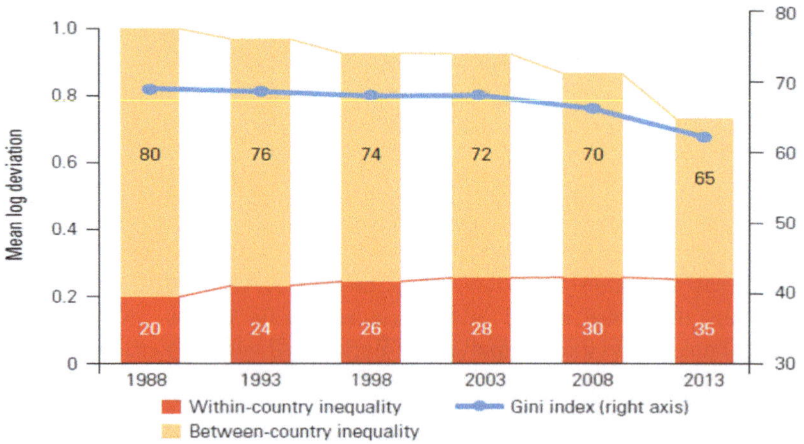

Chart 2 Global inequality, 1988–2013. *Source* World Bank (2016). Poverty and prosperity 2016: taking on inequality

corresponded to 40 per cent of the sectoral productivity growth. This is substantial after a decade in which that contribution reached 0.4 percentage point a year.

And before anyone thinks these results are only due to China, they are robust even when one excludes China from the analysis. China is, of course, a unique case because of its size and growth rates. But the fact of the matter is that this is an observation that can be generalized about the transfer of knowledge.

Of course, this contribution of foreign knowledge translates itself into higher results when accompanied by domestic endeavours. As our colleagues at the World Bank have highlighted in many studies, there is a component of technological capabilities that is idiosyncratic and local. Capabilities must be present in order to utilize foreign knowledge effectively. This has been the case for countries like South Korea and China, as evidenced by their patent filings and R&D expenditure.

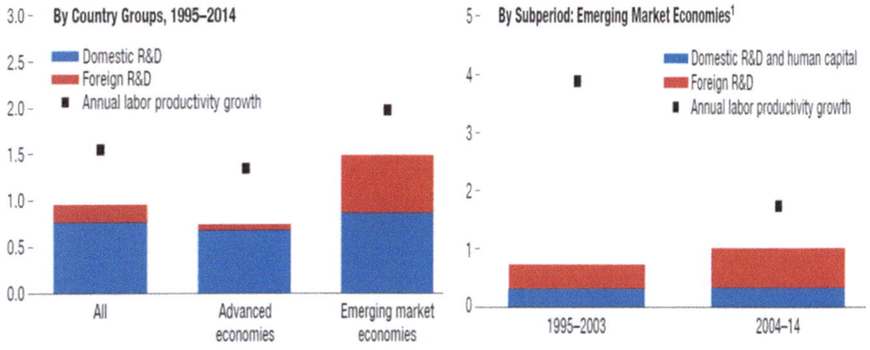

Chart 3 Contribution of foreign knowledge to labor productivity growth annual percent growth, cross-country averages. *Source* IMF. World Economic Outlook, April 2018

Okay, so then we delve into the phase of slowbalization. Looking in a bit more detail, we note in Chart 4 that the cross-border flows of goods, services, and capital slowed after the global financial crisis. There are several hypotheses about this. One of them is the possibility that the major wave of fragmentation associated with manufacturing had reached a plateau. For it to continue as a driving force, we would need to see what happened in China replicated in other countries. This started to happen slightly with countries like Vietnam. India remains the significant absentee in this process.

Another hypothesis is that advanced countries transitioned more towards service-based economies. Services are less trade-intensive, and the internationalization of services hasn't occurred to the same degree as we've seen with manufacturing.

It is important to highlight, as I said, that the average industrialized country saw an increase in the Gini Index from 30 to 33 in the 20 years between 1988 and 2008.

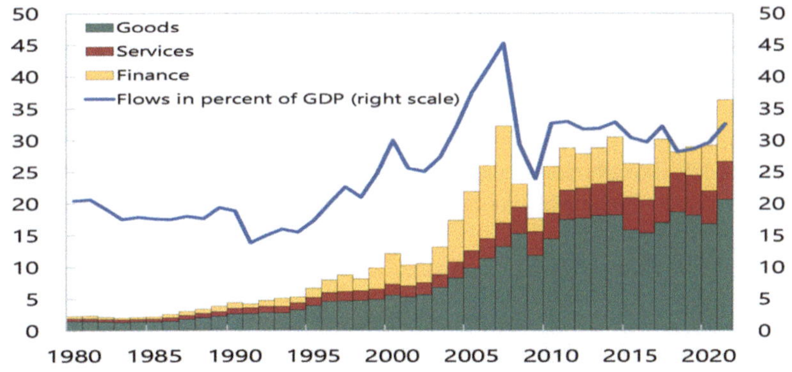

Chart 4 Global flows of goods, services, and finance (US$ trillion, unless indicated otherwise). *Source* Aiyar, S. et al. (2023). Geoeconomic fragmentation and the future of multilateralism, IMF staff discussion notes SDN/2023/001

Now, to avoid any misunderstanding, it must be clear that globalization cannot be held mostly responsible for the rise in economic inequality in advanced economies.

As it was well remarked in our previous session, technological change had more to do with that. Technological change, combined with a lack of appropriate social protection systems in some advanced economies, is to blame for most of worsening of job and income conditions at the bottom of pyramids in countries like the U.S. or the United Kingdom. Globalization cannot be scapegoated despite the claims blaming imports of goods from Mexico and China as responsible for doldrums faced by low- and middle-income workers in the U.S., or blaming labour immigration as a motivation for the Brexit decision. The fact of the matter is that globalization cannot be held responsible for that.

Then the global economy went through the recent multiple shocks, the perfect-storm combination of a pandemic, war in Ukraine, manifestations of climate change, the emergence of the so-called 'new Washington consensus,' and the ongoing technological rivalry.

Let me touch on the impacts of those shocks. The permanent impacts of the pandemic will be limited. The pandemic introduced a trade-off between resilience and efficiency. But the fact of the matter is that this doesn't lead to reshoring. As the milk-formula experience last year in the United States showed clearly, if you bring back everything, then you'll remain as exposed to potential risks as if you were when relying on global supply chains, given the possibility of shocks at home. On the other hand, depending on the sector, this logic will lead maybe to some costly diversification or duplication of links depending on the sectors, but not a reversal of globalization. As some colleagues and I have shown in a policy brief for the T20 this year, the recovery of manufacturing output, particularly in technology sectors, was really nothing commensurate with the stigma established with the pandemic.

Now, where the danger lies is in the rise of national security commanding economic policies, as it has been singled out as a justification for trade restrictions in those sectors where 'dual use' of technologies and goods and services for both civil and military reasons is possible. Indeed, if one looks at trade and FDI restrictions, the rise has been unequivocal, often justified based on national security reasons.

The transmission channels of the fragmentation will be a reversal of the path along which we have attained the gains that we approached before. As we are at the beginning of this process, any estimate of the costs is based on simulations on different models. For illustration, Chart 5 displays results of some studies presented in a recent seminar at the IMF on several models coping with different aspects of the process of trade fracturing.

One can generalize the following from those studies:

1. The costs are greater the deeper the fragmentation.
2. Reduced knowledge diffusion due to technological decoupling is a powerful negative amplifier of the trade channel.
3. Emerging markets and low-income countries are most at risk from trade and technology fragmentation.
4. Transition costs can be considerable, in some cases even exceeding the final trading impact.

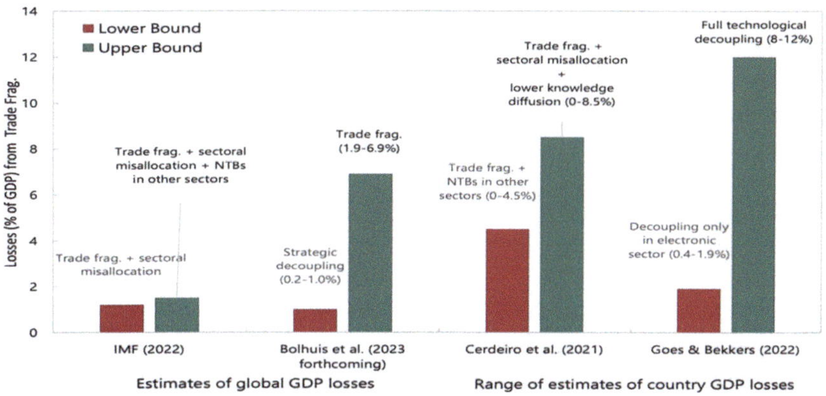

Chart 5 Long-term losses from global trade fragmentation (percent of GDP). *Source* Aiyar, S. et al. (2023). Geoeconomic fragmentation and the future of multilateralism, IMF staff discussion notes SDN/2023/001

5. The estimates provided are not the upper bound.
6. To finalize, the G20 might not address issues of national security directly, but there's much they can do, especially regarding the trade-offs between resilience and efficiency, designing policies to avoid resorting to the least discretionary breadth.

Thank you!

3.5 Expert Comment: B. V. R. Subrahmanyam

I would like to make a statement firstly, as a bureaucrat. And it's only about myself as a bureaucrat. I'm humble enough to believe that I can't pick winners. If I could have picked winners, I would've been a businessman myself, and there are enough people in government who think like me. So, I think that answers a lot of questions.

You have to realise it's largely for two objectives. Firstly, there were huge internal disadvantages within India and it was largely to overcome bad logistics, bad electricity, supplies, etc. The second one was to promote production in areas of strategic importance.

Having said that, there were three areas I could have spoken on. I could have spoken on supply chains, India and India's stance on the trade of the WTO. But since the least amount of discussion has been on the WTO, let me say a couple of things. The topic was the fractured trading system. And we have heard in the last hour or so about the risk of fragmentation, protectionism rising, nearshoring, French shoring, populism on the rise etc. The fundamental question is why is all this happening? There are multiple reasons.

One, I think the WTO has not delivered, and it has not delivered for a very long time. After the round, what else has it delivered in terms of substantive liberalisation, maybe the IT agreement, of which India has benefited a lot. There is a lot of impatience in the world with respect to the WTO and developed countries, there are no more tariffs to bring to the table to extract concessions. So, there is also a loss of patience there. There is no movement on agriculture. I think the developing countries are also quite fed up with that. And the services agreement is too complex to negotiate. I mean, you have to negotiate with 160 countries or 80 countries in parallel. It's just not a workable system.

And that is why there is a gradual increase in behind the border issues, and the WTO system has been jammed in terms of being unable to tackle the labour environment and a whole host of things. And then also these supply chain issues, concentration that has also led to fragmentation.

We have heard about geopolitics, but then I would like to take you to some trends that are happening, which are going to overwhelm. These are important to understand why the system will adjust whether we like it or not. At the end of the day, systems are not permanent. They respond to situations and if the existing ones don't work, we have to come up with new ones just like the G20 has become a more important forum.

It was a Finance Ministers' association, and then it became a Heads of Government forum. It is addressing issues of which finance is just a small part. The trends are on the trade front, the rapid growth in services, it is so fast. In India's context, it is going ahead of manufacturing very soon in terms of trade in services. And the second is the mixing of our services with goods in any exported goods product. Today, 30% to 50% of the value is in its services. So actually, that is a megatrend, which we all have to recognise. The second is demographics. Demographics are going to lead to shortages around the world of labour. I think that needs to be looked at.

And the other realisation, and I'm saying this applies across the world, that behind the border issues are no longer treated as such. They are actually trade issues. And I think, for countries like India, which used to take a position that these are behind the border issues, we will not touch them, are not taking this stance anymore. We have changed our position because it is a gradual realisation that we need to evolve with the rest of the world. And so, India today, when we are negotiating an FTA with the EU, UK and others, we are discussing environment, labour, gender transparency, anti-corruption and what not. Ten years ago, five years ago, India would have been a solid opponent of all of these at the WTO.

One of the first things I had to do as Commerce Secretary was to remove all the old WTO staff in the ministry and change them because you have got people stuck in old positions and they have made their life actually selling those positions. You need new people to do that. The last is an important megatrend i.e., the massive flow of intellectual property across borders, the massive flow of technologies across borders, the massive flow of students across borders. I mean, this is going two to three times faster than the growth of trade. Now, all these things also countries need to regulate. They need to talk and they are talking bilaterally in agreements, in communiques.

As I see it, the multilateral system is not delivering. Very deep, either bilateral or regional trading agreements, will go into a whole host of issues the WTO currently is unable to deliver on. That does not mean that the WTO will not deliver. What will happen, particularly for developed countries which either do not have the capacity to negotiate in these areas, or which actually do not have past experience in negotiating in these areas, is that they will actually test the waters bilaterally or regionally. Then the same issues will arise multilaterally.

For the first time after 20 years, the WTO entered into an agreement on fisheries. It can add 20 years down the road, an agreement on labour, an agreement on the environment, an agreement in many other areas, which are inevitable. I am actually willing to put my money on it. It is going to happen because after you assign these things bilaterally, plurilaterally with a smaller group of countries, how can you resist the argument at a multilateral level?

Only thing is everybody has to get on board. WTO has a peculiar structure because of which it becomes difficult to get things into it, but I think it is not far away. After all, what are regional trading agreements? Mega regional trading agreements? Just that they are not implemented at the WTO and they are not part of it, but they are in a way de-facto pluri-lateral agreements. Without having the WTO's dispute settlement mechanism in the agreements, they have a separate dispute settlement mechanism.

So, I think the system is moving, it is evolving, and I see that India's moves in the last two years have been in that direction. India negotiating with the EU, with the UK and all others is actually a stepping stone. India will have to create a role at the WTO. Actually, what has happened, I mean, with respect to India changing its stance, it cannot stand in splendid isolation. It is the only major economy. It is not part of any significant regional grid. And it is also losing out because of this, all the inherent advantages it has to export.

And I think that is something which is gradually been realised and my own personal feeling without having data is that not being integrated with large parts of the world economy in many areas is going to hurt India, also in greening itself. And I think that is very, very essential.

The other point which I heard is, this question about, how do you separate out what is being done for political reasons? What is being done for other reasons? As a bureaucrat you have to take sociopolitical systems as given, and there is no point in trying to avoid them.

The best deals are those which actually manage to sell things to the socio-political system. It may not be economically the most perfect answer, but if you pay the economic, the political and the social cost, you can get the deal through. I think an 80% or 50% deal is better than no deal.

And it works. I think this is the way forward. In the Indian context, the country has changed its stance, but countries do change their stance. It is like steering an aircraft carrier, you know, it takes a very long time to turn, but when it turns, it turns and then it sticks to it. So, I am very hopeful. And, for all this fragmentation etc., which you see, I anticipate the global system evolving to accommodate it.

References

1. Mattoo A, Mishra P, Subramanian A (2012) Spillover Effects of exchange rates: a study of Renminbi, Working Paper 12–4, Peterson Institute for International Economics, Washington DC
2. Subramanian A, Kessler M (2014) The hyperglobalization of trade and its future. F. Oxford University Press, Allen et al, Towards a better global economy, Oxford, pp 216–277
3. Amirapu A, Subramanian A (2015) Manufacturing or services? An Indian illustration of a development Dilemma, WP 409. Centre for Global Development, Washington DC
4. Kumar, Nagesh (2018) 'Reversing the Pre-Mature De-industrialization for Job-Creation: Lessons for 'Make-in-India' from Experiences of Industrialized and East Asian Countries', in A. Ghosh Dastidar et al eds. *Economic Theory and Policy amidst Global Discontent: Essays in Honour of Deepak Nayyar*, New Delhi: Routledge, 389–415
5. Rodrik Dani (2015) Premature Deindustrialization, WP #20935, National Bureau of Economic Research
6. Kaldor N (1967) Strategic factors in economic development. New York State School of Industrial and Labour Relations, Cornell University, Ithaca, NY
7. Nayyar D (2019) Resurgent Asia: diversity in development. Oxford University Press for UNU/WIDER, Oxford and New Delhi
8. Rodrik Dani (2022) The new productivism paradigm, Project Syndicate. https://www.project-syndicate.org/commentary/new-productivism-economic-policy-paradigm-by-dani-rodrik-2022-07
9. The Economist (2022) Many countries are seeing a revival of industrial policy: A previously discredited approach has found new believers, *The Economist, January 10th 2022*
10. Juhasz, Rek, Nathan Lane, Dani Rodrik (2023) *The New Economics of Industrial Policy,* NBER Working Paper 31538. http://www.nber.org/papers/w31538
11. Siripurapu Anshu, Noah Berman (2022) Is industrial policy making a comeback? Council for Foreign Affairs. https://www.cfr.org/backgrounder/industrial-policy-making-comeback
12. Chen Ellie (2023) CBAM: Climate Change Savior or Protectionist Ploy? *Chicago Policy Review,* April 17th, 2023 https://chicagopolicyreview.org/2023/04/17/cbam-climate-change-savior-or-protectionist-ploy/
13. Kumar N (2002) Globalization and the quality of foreign direct investment. Oxford University Press, New Delhi
14. Kumar, Nagesh (2020) East Asia's Paths to Industrialization and Prosperity: Lessons for India and Other Latecomers in South Asia, *Economic and Political Weekly,* LV(50), December 19: 24–31
15. Kumar, Nagesh (2022a) Indian Economy @75: Achievements, Gaps, and Aspirations for the Indian Centenary. Indian Econ J 70(3). https://doi.org/10.1177/00194662221105552
16. Kumar N (2017) National development banks and sustainable infrastructure in South Asia, GEGI WP 003. Pardee School of Global Studies, Boston University, Global Economic Governance Initiative
17. Nayyar Deepak (2015) Birth, life and death of development finance institutions in India. Econ Pol Weekly L(33):51–60
18. Arun Kumar, Ramaa (2023) Impact of Import Liberalisation on Employment in Manufacturing Industries in India, ISID Working Paper No. 265
19. Chang, Ha-Joon, Kiryl Zach (2019) 'Industrialization and Development' in Nayyar, Deepak ed. *Asian Transformations: An Inquiry into the Development of Nations,* Oxford: Oxford University Press
20. Lin Justin Yifu (2019) 'China' in Nayyar, Deepak ed. *Asian Transformations: An Inquiry into the Development of Nations,* Oxford: Oxford University Press
21. NITI Aayog (2023) National Multidimensional Poverty Index 2023. https://niti.gov.in/sites/default/files/2023-07/National-Multidimentional-Poverty-Index-2023-Final-17th-July.pdf
22. Cheriff, Reda, Fuad Hasanov (2019) The Return of the Policy that Shall not be Names: Principles of Industrial Policy, IMF WP/19/74

23. *The Economist* (2019) 'Slowbalisation: The steam has gone out of globalization,' 24 Jan 2019. https://www.economist.com/leaders/2019/01/24/the-steam-has-gone-out-of-globalisation
24. Franklin Foer (2023) The New Washington Consensus. *The Atlantic,* https://www.theatlantic.com/ideas/archive/2023/05/biden-economics-industrial-policy-trump-nationalism/673988/
25. ISID (2022) Financing India's Industrial Transformation: Some Policy Lessons from International and National Experiences, New Delhi: Institute for Studies in Industrial Development
26. Kumar Nagesh (2022b) Competitive manufacturing as a driver of India' next economic transformation: opportunities, Potential and Policies, New Delhi: ISID WP#259
27. Kumar Nagesh (2023a) Unlocking India's potential in industrial revolution 4.0: national innovation system, demography, and inclusive development. Indian Public Policy Rev. 4(3):67–87
28. Kumar Nagesh (2023b) India's evolving industrial policy is critical for realizing its development vision. *ProMkt,* Spring: 54–59, Stigler Center, Chicago Booth

Chapter 4
Reshaping Global Finance for Sustainable Growth

N. K. Singh, Hanan Morsy, Tao Zhang, Poonam Gupta, and Manjeev Singh Puri

4.1 Session Chair: N. K. Singh

The theme today for this session is reshaping global finance for sustainable growth. To some extent, there is a degree of ambiguity in the formulation of the broad theme of this question. Of course, the reform of the global financial architecture is an issue with which, in some ways, as you said, I'm grappling with currently. It's not an easy grapple. The global financial architecture is cluttered with all kinds of other non-economic issues, which have come in display. Broadly, I will say that the global financial architecture is dysfunctional and requires a basic, not rejuvenation, but a fundamental restructuring to meet the needs of adequate finance for an orderly transition to a green economy.

Sheer politics and other kinds of considerations have begun to play an important role in deflecting us away. There is clearly an inadequacy of finance. The kind of financial requirement is anybody's guess, I mean, I was audacious enough to pick on an old figure that, from the sort of figures currently in use in the United Nations in the OECD, in the IMF and the World Bank, they believe that you roughly require

N. K. Singh (✉)
Chairman, Finance Commission and President, Institute of Economic Growth, New Delhi, India

H. Morsy
Deputy Executive Secretary and Chief Economist, United Nations Economics Commission for Africa, Addis Ababa, Ethiopia

T. Zhang
Chief Representative for Asia and the Pacific, Bank of International Settlements (BIS), Hong Kong, China

P. Gupta
Director General, National Council of Applied Economic Research (NCAER), New Delhi, India

M. S. Puri
Former Ambassador of India to the EU, Distinguished Fellow, The Energy and Resources Institute, New Delhi, India

S. Bery et al. (eds.), *Navigating Challenges for Sustainable Growth*,
https://doi.org/10.1007/978-981-97-7894-2_4

about $3 trillion a year between now and 2030. If you ask me whether it could be 2.8, I will say could be, if you ask me, could it be 3.2, I'll say could be, and I have no better answer to give, except that it has to be significantly, enormously higher than the current capital and the current investments which are going in.

So having gone into the requirement of that kind, how is this requirement going to be in some way met is the principal issue now. The assumptions which are being made is that two out of this three trillion should come from domestic resource mobilisation. That's easier said than done. In fact, Christina, the managing director of the IMF, told me just last week that you know, "NK, everybody's talking about your $1 trillion. Nobody's talking about that $2 trillion." I said, "Well, Christina, better you talk about that $2 trillion." Because the IMF is in some ways positioned to do that given that they are looking at revenue, revenue buoyancy, tax-GDP ratio and the kinds of domestic reforms needed for economies to be able to garner that kind of capital.

So that leaves $1 trillion. Where do I get $1 trillion from? $500 billion should come from private capital. Is it possible? I don't know. What will make it possible is the issue with which we are dealing with, and then the balance, $500 billion, a mix of concessional and non-concessional capital.

I did not realise that there was a separate MDB for Black Sea development until it was brought to my notice. There is a whole family of MDBs, and of course the World Bank Group, which is the IBRD and the IDA, and others (International Finance Corporation, MIGA, Global Environment Facility), all added together their annual lending as of 2019 was between $110 billion to about $120 billion.

The challenge is how do I triple that $110–120 billion, and then you add private capital to get to $1 trillion. Then in some form, that's one way of addressing the compelling needs of what you call global reshaping. Global finance means most importantly availability of the adequacy of finance to address the needs of sustainable growth.

By the way, let me add that Africa is one of the very dominant and important areas of focus as I grapple with the problem. It is generally known but not many seem to address it, is that it is one continent where in many parts extreme poverty has gone up, shared prosperity has gone down. I must say in a lighter vein, it was told to me very often, "Look here, don't tell us about all this green and non-green. Before we begin to make choices on energy and exercise energy options, we do not have any energy." So, these choices only become meaningful when you at least address the basic energy availability.

4.2 Speaker 1: Hanan Morsy

Africa has had a developmental progression in terms of all the achievements that have been made on SDGs due to the tsunami of global shocks. While we discuss climate action and the necessary steps, at the same time, Africa has three-quarters of its population without access to electricity.

Let me give some background. Part of the work I do involves heading the secretariat for the Africa High-Level Working Group on Global Financial Architecture. This group is composed of African ministers of finance, economic development, and planning, along with the IMF, World Bank, African Union, African Development Bank, and African Bank. The group's aim is to build consensus on the African position regarding the reforms required for the global financial architecture. This group was formed at the beginning of last year, 2022, and has been meeting regularly to develop these asks.

One of the key components discussed relates to the IMF's special drawing rights (SDRs) and its role in the global financial safety net. For background, the SDRs were established in the late 1960s to supplement official reserves and facilitate global liquidity. According to the IMF's Articles of Agreement, there are two instances when allocations should be considered: every five years and during unexpected major developments. Out of the 12 periods since the institution of SDRs, there have been only four general allocations and one special allocation. You can see, even in terms of size, there are two significant allocations in 2009 after the global financial crisis and then in 2021 following the pandemic. What's crucial to observe is that even when we have the SDR allocations, which are intended to assist with global liquidity issues, their distribution is based on IMF quotas, favouring countries with larger economic sizes and better financial positions. Essentially, these allocations often end up with countries that don't necessarily need them the most.

Furthermore, when examining utilisation across developing versus developed countries, developed countries, despite receiving the majority of the allocation, utilise less than 6%. In contrast, developing countries utilise over 40%.

Several recommendations have emerged to reform this system:

i. Make the allocation process more rule- based, analytical, and less discretionary.
ii. Introduce clear and automatic triggers, such as force majeure allocations linked to pandemics and natural disasters.
iii. Consider defining a global recession as two consecutive quarters of negative growth, as commonly defined by economists.
iv. Recognise widespread capital flight from emerging and developing markets as one of the triggers for allocation.

In refining the system of unexpected and five- year basic period allocations, it's essential to consider how to reform the allocation formula. The aim should be to ensure that the resources go where they are most needed, true to the original intent. This means, in addition to the IMF quota consideration, the formula should include liquidity considerations, ensuring that countries in genuine need receive more liquidity.

And then the other issue which we've been stuck with following the last allocation is how do we really stimulate the SDR re-channelling? So, following the last SDR allocation, a number of countries have pledged that they would re- channel these SDRs so that countries that needed more can benefit. But there have been a number of obstacles. Some of the things that can be done to actually unleash that is, for example, to enable re-channelling of SDRs to MDBs. And we actually have a viable option that has been put on the table by inter-American Development Bank and African Development Bank that need five SDR donor countries to make it operationalised.

And of course, there are other things also in terms of reviewing the SDR reserve asset characteristics to make it more up to date with the practice of many of the major central banks.

And then on the other major area that we've been working on are reforms to the global debt architecture. Here, this gives you a sense of the evolution of the debt landscape for Africa, and you can see that things have been changing, and part of it is driven by the fact that there has been a decline of official development assistance and availability of concessional financing, which also contributed to having a much higher share of commercial and private creditors.

But also, there has been an evolution of the landscape of official creditors with a more increased role of China and other creditors rather than just the Paris Club. This has implications for the issue of debt resolution and restructuring. Some of the recommendations that came out of the working group relate to the need for overhauling the G20's Common Framework. For instance, we had African countries that, from the time they applied until a creditor committee was formed, took two years. This illustrates how the process has been very slow.

There's a pressing need to reform it so that it's more time-bound, efficient, and transparent. Implementing a suspension of debt service upon application and bolder use of IMF lending into arrears for both official and private creditors is essential. There's also a need to establish a comparability of treatment formula to reduce technical disputes and expand creditor committees to include private sector creditors.

Other significant areas of the global debt architecture that require attention include the regulatory side. There's a need to enforce enhanced collective action clauses in debt issuances and contracts and introduce force majeure and climate-resilient debt clauses. We had some progress this year in this area, and there's also a push to implement vulture fund legislation. In the long term, there seems to be a need to establish a multilateral creditor club to coordinate the framework and oversee the outstanding global debt issues.

Another pillar has been related to market access. One prominent issue is the Africa Premium. African countries tend to pay 150 to 250 basis points higher than countries with the same economic fundamentals. This discrepancy is primarily driven by two factors.

Firstly, there are information asymmetries. There's a call for support from the G20 to enhance this area by building capacity and increasing resources. Secondly, expanding partnerships to improve market access, particularly in areas like ESG investment and green capital markets, is vital.

Another critical aspect is the regulatory side of credit rating agencies. There's a perceived bias in ratings, even for countries with similar economic fundamentals. This highlights the need for oversight in the industry to ensure fairness and accuracy.

We recognise the oversight measures established in the European Union, but there's a pressing need to consider these on a global scale. This involves a thorough examination of the current regulatory framework and proposed reforms to improve oversight.

And there are a number of them that we can talk about in terms of the green finance area. Despite that, Africa emits less than 3% of global emissions, and it's the most vulnerable region to climate change. Africa benefits less than 1%, has basically less than 1% of global green bond issuances. And as we talked earlier, it tends to be higher cost. So, there is really a need to tackle these issues.

From the IMF side, I think the move for having the IMF resilience and sustainability trust is a move in the right direction in terms of having facilities that look like they are more long-term geared and structural and climate action geared, but there is a need to increase the flexibility and eligibility criteria for it and to fast track operationalisation.

Other very important things for Africa in that side is also to strengthen the inclusion of these climate contingency clauses that we've talked earlier about. And to stimulate debt for climate and debt for nature swaps, and to use more guarantees and availability of them to reduce the cost of finance for the continent.

And of course, supporting more de-risking and blended finance because we need to crowd in the private sector much more. The private sector currently is 14%. This is like almost a third of what it is in Asia now. So let me just very quickly say that we have also discussed issues of reforming the Brettonwood institutions.

The way it operates has been more geared toward country-specific shocks and things have evolved with us facing multiple and tsunami of global shocks. So, the tools and the way these institutions have been operating need to change.

4.3 Speaker 2: Tao Zhang

My main message is that there's absolute urgency to find and implement solutions to promote green and sustainable growth. In doing so, we have to give the importance of the market-based incentives for the net zero transition and what can be done by, for example, central banks, which the BIS represents. This is the main focus of my remarks today.

First, of course, you can see that we run a large CO_2 gap between net zero ambitions and the achievement we have made so far. For example, we are seven years into the COP 21 and the objectives of 45% of reductions in GHG emissions by 2030 from the 2010 level is still out of reach. Without further actions of course, we are heading to a 2.8 degrees temperature increase by the end of the century. So, time is ticking.

Second, as many have already mentioned, there's a large financing gap. More or less worldwide, we have averaged 480 billion dollars annually during the last 10 years, which still falls short of the 4.3 trillion dollars annually. The number is quite phenomenal. But the question is, where will the money come from? MDBs, SDR? Some of these are already tapped. Or private sector? But we know that all money comes with a cost and we must have incentives to overcome these costs. For private sector, if they want to put money in, they need a return. The cost is upfront. The benefits come out, who knows, some years later. So how to bridge the gap?

So here I propose a three-pillar approach to accelerate the transition. And the core of it is market-based incentives. This is the first pillar. Why? A moment ago, some people mentioned that who picks the winner, etc. So instead of indicating who should reduce emissions, who should finance the emissions, let the market play the role. And of course, the key is to get the carbon pricing right. The carbon pricing, of course, sends economic signals to emitters and allows them to decide to either transform their activities or lower their emissions or continue the emission but with a higher price. So that's as simple as that because otherwise we just end up debating who is picking up the thing. And of course, right now, carbon pricing can take all kinds of forms, for example, emissions trading systems or carbon crediting mechanism, to name a few. But we have to admit the progress has been very slow. And we have to ask ourselves why is that?

Earlier this morning, I recalled some participants mentioned that we would need to promote the global emission trading system. I wholeheartedly support it. But between now and then, we have a lot of work to do. And so far, by the end of 2022, there are only 34 ETS globally, covering only 17% of global GHG emissions. And carbon credit market is only voluntary for business. So there must be some huge impediment on it. We have to work on it, find out the solutions, and I think the G20 can work towards that.

Now, the second pillar. Of course, as everyone can imagine, that we need international cooperation as no one can achieve the climate objective alone. And for obvious reasons, high emitting industries will simply migrate to where the climate regulations are the loosest. So, unless all jurisdictions act in a closely coordinated way, these things cannot be done. And of course, international cooperation is necessary also on the account that there is a great need to ensure the equitable burden sharing between developed and developing countries in pursuing the net zero objectives.

And the reason is very simple and compelling. Developing countries are hit hardest by climate change, but they are also the least able to finance climate mitigations and adaptation. So external financing and technological transfers are most needed for developing countries to build clean and climate-resilient features. The question is how to make it happen. This is up to, of course, G20 to answer.

So, the third pillar, we have to rely on public policies. So here, to get the carbon pricing right, market forces sometimes cannot achieve it alone. Public policies play an important role in it. For example, carbon tax needs a base: how much and how broad. And of course, this also relates to the removal of energy subsidies. And of course, related to the associated regulatory policies that also impose the shadow price on pollution. So in general, public policies play an important role. Here, because I come from central banks, I have to say a few words on central banks, particularly, the BIS, what our role could be. Very briefly, three roles we can play.

First, as green investors. Indeed, many central banks, including us, the BIS, have already adopted the investment strategy. The ECB is taking the lead on this front, and indeed, India earlier this year also issued a green bond. And the Indian government also issued the framework which highlights the qualifications and disclosure requirement for green bond. So things are happening on the ground. For us, the BIS has set up three Green Bond Funds. The number is around 3.5 billion USD, not huge compared to the trillion-dollar requirement. But we hope, with our issuance though, the private sectors and others can follow up.

Second, we can also work as a technology enabler. In a sense, the central bank can test promising new technologies for green finances. And here I would like to do a little bit of promotion. Five years ago, the BIS set up something called the Innovation Hub at the headquarters. Subsequently, we set up nine Innovation Hub centers across the world, including two in Asia, one in Singapore and the other one in Hong Kong. In these centers, we are piloting a number of projects utilizing new technology, including blockchains, to ensure that green financing can benefit from the new technologies, improving efficiencies and impact measuring.

And third, as regulators, central banks can regulate and implement disclosure measures, and set regulations for the banks, working together with other standard-setters like FSB and BCBS. To conclude, there's real urgency among us to act now and act early. Asia should take the biggest benefit from it because we are not only the most populous region in the world, but also the most dynamic in terms of economic activities.

4.4 Speaker 3: Poonam Gupta

My presentation is based on my past and ongoing research (for one decade each at the IMF, and at the World Bank, and the last two years at NCAER, India's leading economic policy think tank).

As has been said at the conference today throughout the day, financing needs have increased around the world. Meanwhile, the global liquidity has become scarcer and private capital flows have remained as fickle as ever.

Three kinds of financing are available to the emerging market and low-income countries. First, the private sector funding at market rates, which is volatile and can flow in and out at very short notices. Such funding is available only to the emerging markets which are able to attract private sector funding.

Second, the multilateral funding, which is long-term and concessional, but is limited in volume. Increasingly, it is the low-income countries that have greater access to multilateral funding.

Finally, there is bilateral funding, which is smaller in volume compared to the first two, and has increasingly become more strategically tied with specific projects. In bilateral funding, new funders, particularly China, have replaced the traditional funders.

When we talk about global finance for sustainable growth, we need to de-risk private capital flows for emerging markets, as well as rethink the financing envelope and sources for low-income countries.

The capital flows to emerging markets have become more volatile over time, and the episodes of reversals have become more frequent. The nature of such reversals has changed as well. Earlier the reversals would occur as country-specific sudden stop events, e.g. those in Mexico 1994, and the East Asian countries in the late 1990s.

Increasingly, these events have been occurring as emerging markets-wide sell-off events, when capital flows suddenly reverse out of emerging economies back to the advanced economies. Recent examples of these episodes include the first taper event in 2013, and the subsequent ones in 2016, 2018, 2020, and 2022.

When the reversals of capital flows happen, they create volatility, and thereby challenges for the policymakers. The latter, within a very short period of time, have to manage their exchange rates, communications, market sentiment, and the levels of foreign reserves.

These reversals and their impacts accrue despite the fact that emerging markets now maintain strong economic frameworks, specifically the kind of frameworks that the IMF recommends. These countries have more resilient growth outlooks, strong fiscal positions, credible fiscal and monetary policy frameworks, and independent central banks. Besides, they hold large volumes of foreign reserves, limit the exposure to foreign currency in their debt portfolios, and maintain flexible exchange rates.

India is one such example. Despite its strong policy frameworks and a resilient economy, it too gets subjected to the reversals of capital flows from emerging markets during an emerging market sell-off episode.

A key question to ask is: How do we make access to capital flows safer for emerging markets?

For this, the global financial safety nets ought to be strengthened, and this is where the role of multilateral institutions such as the G20 and IMF and bilateral arrangements becomes important.

Countries generally sign up for bilateral swap lines in hard currency liquidity, and often with many countries at any given point in time. But these have not been enough to insulate them.

The bilateral swap arrangements have proliferated during the last decade, led by China. China has offered swap arrangements to many low-income countries, with countries that it considers of strategic interests, and with ones that it has had strong trade relationships with.

The US' Federal Reserve Board too offers its own swap lines, but in a rather limited and selective fashion. It has offered them to other fellow advanced economies and only to four emerging markets, which do not include India. Empirical results show that the bilateral swap arrangements are of limited use unless these are from a large, credible country, which issues hard currency, such as the US.

Countries also sign up for regional financing arrangements, e.g. the Chiang Mai, BRICS and the SAARC initiatives. These regional arrangements have mostly not been invoked at times of the capital flow reversals.

Finally, there have been three IMF contingency lines, with the first one introduced in 2009. The latest one, the third one, was established in 2020. Yet, until 2021, only eight countries had ever signed up for those, including Mexico and Columbia, among others; and only three countries have ever drawn funds. Further, there has been limited innovation in these credit lines despite this very limited use.

Incidentally, however, these global financial safety nets have been inadequate or ineffective in making the flow of capital safe and resilient. As a result, the emerging markets are left with few choices but to accumulate and use their foreign reserves more actively.

The credit rating of an emerging G20 country versus an advanced G20 country yields interesting results and a potential area of reform. An average advanced G20 country has an almost perfect rating. If the ratings are converted into numerical ratings ranging from 1 to 20, an average advanced G20 country boasts of a near-perfect rating of 19, while an average emerging G20 country has a rating that's seven and a half points lower, just a notch above the junk rating. This disparity is inexplicable and persists even after accounting for the obvious sources of variation.

When it comes to the financing for low-income countries, a large number of these economies are facing some degree of debt distress. The average general government debt to GDP ratio is high, with a significant portion of it being raised externally. Breaking down the lenders, we find that the multilateral institutions hold about 50% of this debt, even as the new bilateral lenders are emerging, with China leading the pack.

Regarding finance for low-income nations, multilateral development banks, including the World Bank, conduct their debt sustainability analyses. Yet, if most of the low-income countries are mired in debt distress, one may question the efficacy of these analyses and ask whether the debt sustainability analyses ought to be conducted more robustly.

Exchange rate risk is another significant concern that warrants discussion. How can we mitigate, at least partially, the exchange rate risk associated with external financing for low-income countries?

Thus, on rethinking global finance during the G20 presidency, the following lessons emerge.

First, the buildup of foreign exchange reserves by emerging markets should be supported.

Countries shouldn't be labeled as currency manipulators when they utilize their reserves.

Second, the model and governing frameworks of credit rating agencies need to be regulated more fairly.

Third, the G20 should motivate central banks to expand their currency swap networks. Relatedly, the issue related to the spillovers of monetary policy ultimately originates in the US. It may issue the swap lines more widely. Swap lines from other issuers of hard currencies ought to be expanded as well.

Fourth, the IMF ought to extend the use of contingency lines. They can declare in their Article IV reports as to which country qualifies for which one of the lines and for what amount. There should be automatic triggers for when such liquidity becomes available. The countries should have to pay for the lines only when they draw on them.

Finally, we may also better account for the impact of the policies of advanced economies on emerging markets. There has to be a more balanced cycle of monetary expansion and withdrawal. During the financial crisis of 2008–2009 and again during the COVID crisis in 2020, when monetary policy needed to be eased, the issue was discussed extensively in G20. Thereafter, within the G20, all countries agreed to ease their monetary policies. However, when it was time to normalize these policies, the decisions were taken and implemented bilaterally without involving all the member countries. There is obviously a need for greater symmetry in this process.

4.5 Expert Comment: Manjeev Singh Puri

I want to draw your attention to the hard realities of climate change. Let's take India as the benchmark. Where is the most important amount of action required to be taken? It is the large developing countries, most of whom qualify as emerging markets.

Secondly, do they have sufficient capital by themselves? Can we simply squeeze it out of them? As it is normally said in multilateral meetings, we will do best practices sharing, capacity building, tell you how to go about doing it. You really think that's the case? I can give you the figures as far as India is concerned, only looking at the announcements we have made till now on renewable energy, etc. even to meet a part of them would take care of a fourth of all deposits in Indian banks and so on and so forth.

And I'm not talking about adaptation, which is a completely different ballgame. We have no choice but to get capital from overseas. Macroeconomics, swaps, portfolio investments, all those are issues. Why are our companies not able to raise money overseas? Of course, there are some cases which are very successful, but I want to tell you in the context of his report, Mr. N.K. Singh carried out a small group session, which included two of India's large players in the international market, one from the private sector and one from the state sector. And Sir, I hope you will recall that the one word that both of them used was bring down hedging costs.

Do you know what the hedging costs today are? For some of India's largest and best- known groups, they could be upwards of 4%. This is the reality of things. Today, interest rates worldwide are a little high. A few years back they were rock bottom. But what was it that our countries and our companies were able to meet?

Where are we going to be able to get funding? We have no choice but to turn to what is the situation?

Is there a macroeconomic deficit? No, there isn't. In the world. There is enough money. We who need it don't have it. How do we attract this money by making investment in our countries interesting to them, worthwhile for them and for our people who are doing business affordable. There are two sides to this point.

Can we just do one little thing? It's not a panacea, but will it help? And that simple thing is can we do something about foreign exchange costs? I want to read to you something from the triple agenda. Very lovely paragraph, small one. I'll read two lines. A particular pain point for private investments. If private investors is exposure to currency risk. And then they go on to say that they will deal with it in volume two. These are simple words, a pain point. What does it mean? It simply means that on both sides, affordability remains a real problem. Can we do something about it? And you know, I don't want to get in again, as I said, on geopolitics and so on.

The smallest developing countries, the least developed countries, well, there is much that we need to do for them, all of us, including emerging countries. If we can't get things done for ourselves, what are we going to do?

And what are some of these ideas? If you take this kind of agency at the multilateral bank side, and I believe a bank is a better place than the trust fund under the IMF for this because it is in the nature of action for a particular product, a particular investment, a particular company or a project.

I just want to make this one focused point. What can we do to make the movement of capital from the industrialised countries to projects, companies, corporates doing green work, only green climate-related investments in the developing countries. I dare say much of that will flow to the largest developing countries. But remember, climate change is a global problem. What I do here affects you. What you do there affects me. So, it has a global impact.

Chapter 5
Technology, Policy, Jobs; Multilateralism: Geopolitics, Governance, and the Global Commons

Robert Lawrence and Homi Kharas

5.1 Speaker: Robert Lawrence

I wanted to very briefly touch on my topic, which is technology, policy, and jobs. I'd like to talk about the employment dimension of technological changes that we've experienced in the past and that we're likely to see continue in the future. Then to reflect on some policy responses to those, particularly with an emphasis on what they mean for the labor market.

We have undergone a major transformation in our economies since the 1980s. There are three huge shocks which have hit the economy. The first I would characterize as skilled-biased structural change. Because of relatively rapid productivity growth in manufacturing, the share of employment in manufacturing has declined in many countries.

This phenomenon of skill-biased *structural* change, i.e. lower manufacturing employment shares—has led to much more difficulty in developing countries achieving the same levels of manufacturing employment shares as today's advanced economies were able to achieve. This has meant fewer employment opportunities. Manufacturing was a major ladder to inclusion, allowing workers to join the middle class. Those opportunities have diminished. This is not to say that manufacturing can no longer play a role, especially in countries like India. But it is diminished.

We've also seen within industries, a skill-biased *technical* change. Digital technologies have tended to complement the skills of more educated workers and to

R. Lawrence (✉)
Albert L. Williams Professor of International Trade and Investment, Harvard University, Cambridge, USA

H. Kharas
Senior Fellow, Center for Sustainable Development, Brookings Institution, Washington, DC, USA

S. Bery et al. (eds.), *Navigating Challenges for Sustainable Growth*,
https://doi.org/10.1007/978-981-97-7894-2_5

substitute for the skills, especially of mid-level workers. This has resulted in a polarization of the labor market with fewer opportunities for those workers with mid-level skills.

A third dimension, which I believe is not sufficiently appreciated, is that the nature of capital has changed and it is increasingly intangible. Inputs such as software, databases, patents, algorithms, organizational capital, and branding have become much more important. If you look at the US data for manufacturing, these forms of intangible fixed assets today are more important than equipment. Many people talk about automation, but in fact, it's the intellectualization of production that is increasingly driving capital formation.

As we look to the future, there are challenges, especially for less skilled workers and for countries that seek to specialize in labor-intensive products. We are seeing investments in robotics, 3D printing, and augmented manufacturing. Now, we have AI, whose impact is ultimately uncertain, but certainly, the capacity to substitute for routine workers is going to be considerable. These are pressures which now face all countries, especially those who seek to specialize in unskilled intensive manufacturing products.

However, there are also considerable opportunities which new technologies are going to provide. We now have this unbelievable capacity for people throughout the world to access knowledge. We also have the ability to use technologies to save on bricks and mortar. AI could actually work in both ways. As was noted in the general discussion, e-commerce gives us new opportunities for entrepreneurs. We have the ability to do remote work, not only within countries but across borders. Richard Baldwin calls this 'Globotization.' If those who live in developing countries are able to be employed in advanced countries, earn higher wages, and then spend their money on non-traded goods, this can provide opportunities for less skilled workers.

What has been striking is that, as I look at the new policies being implemented, I think they, by and large, are going to reinforce these skilled-biased technical changes. (Though AI may be an exception). We see the emergence of technological rivalries in which countries are vying for global leadership in technologies which tend to work against less skilled workers. Robotics, additive manufacturing, 3D printing, semiconductors, nanotechnology, advanced materials, autonomous vehicles, telecommunication technologies, and advanced pharmaceuticals all are relatively intensive in skilled and educated workers. The geopolitics are playing into these forceful trends which are tending to work against less skilled workers, highlighting the policy challenges for the labor market.

Green growth as well: My reading of the evidence is that, from a macroeconomic perspective, the best estimates we have, suggest that the net labor force effects of the green transition to net zero are quite small. Although politicians like to accentuate the positive and stress the kind of great new jobs that will be available, there is going to be a lot of disruption in the fossil fuel markets and production of products that are intensive in emissions. If you simply think through the differences in the technologies required, for instance: electric vehicles, which are estimated to have something like 200 parts versus the traditional combustion engine-driven automobile with 1,200 parts, you can predict that there will be a need to move production workers out of automobile production. Today's electric vehicles, in addition to their batteries, are basically computers on wheels.

And so, this is a tremendously skilled-biased technical change that is going to characterize this particular dimension of the Green Revolution. But you can go through other technologies as well, and you can see that while they will create a lot of construction jobs initially, which leads to more inclusive growth, but at the end of the day, the operators and designers are going to strengthen the tendency towards this bias in skill.

We have a lot of people who depend on fossil fuel production. Many of those workers are in far-flung places because that's where we've located the coal mines and other production facilities. The production of renewables will not necessarily be appropriately located in the same places. So, we face a great challenge in adjusting, and there's a lot of talk about 'just transitions' and I couldn't underscore their importance more.

But nonetheless, thus far, my reading of both what the United States has done in its Inflation Reduction Act and what other countries are doing to think about not just the winners but the losers, is totally inadequate to the task of aiding those who will lose. And the same kind of division occurs at the global level between countries that produce fossil fuels for export, oil-exporting countries on the one hand, and the new countries who do have opportunities to specialize in green technologies.

So, if we take our lessons from what happened in the past, the failure to deal with the adjustment problems of the globalization that we have experienced ultimately had seismic political effects. And we saw the emergence of populism and an antagonism to free trade.

Well, we're facing a very similar challenge today when it comes to green technologies, and I don't think there is sufficient emphasis in the discussions on the kinds of policies that are required, both in developing countries and in developed countries, to deal with this dimension of decarbonization. I would underscore the need for more effective transfer programs, adjustment programs, training policies, and the emphasis on skills, given the forces that are working on the labor market today.

And finally, I would simply underscore that whatever the microeconomic policies are, nothing is better for a well-functioning labor market and for creating opportunities than macroeconomic stability. We've been reminded numerous times and seen the evidence of how worker power and worker rights can be reinforced in economies that are able to sustain full employment. And for those at the bottom who are going to experience a lot of the difficulties, the best policies are to drive the economy towards a non-inflationary state of much fuller employment.

5.2 Speaker: Homi Kharas

I'm going to talk about very big themes: geopolitics, governance, and the global commons as they relate to multilateralism. And I'm actually going to do it in reverse order. I'm going to start with the global commons. I'm going to say a few words about governance and then about the politics. Finally, I'll comment on, what does all this mean for a new multilateralism? Because everybody says that if we've got these issues of the global commons, then that requires multilateral cooperation. Well, what does that really mean in practice?

So, I wanted to start with the global commons because it's really the global commons issue that is driving a lot of the change that we see in a lot of the discussion about multilateralism. And the big point about the global commons issue is that it is urgent and it requires a large amount of money to address.

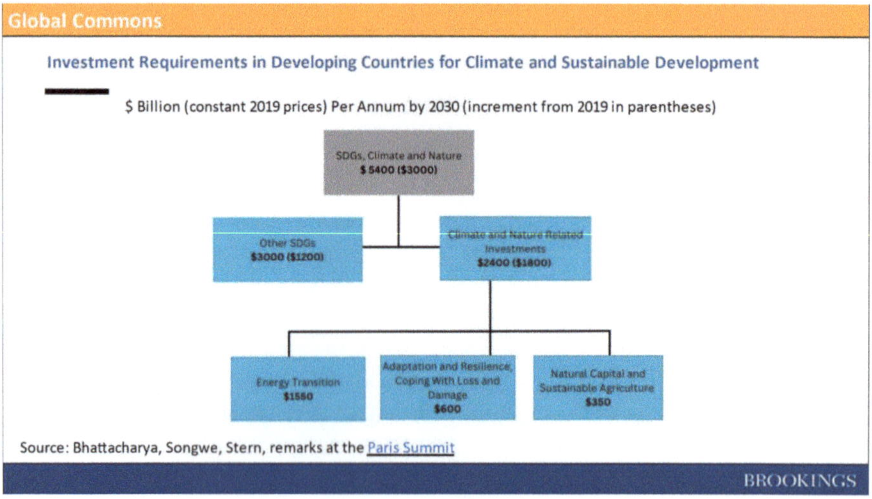

I want to start with some numbers. These are the numbers that you'll see in the independent expert group report for the G20 on strengthening multilateral development banks. You'll also see these numbers now in a whole range of other multilateral discussions. But basically, what we're saying is that developing countries—and now I'm excluding China from this group in these numbers because its reliance on multilaterals is of a different nature, should I say, than many of the other countries— will have to spend something like $5.4 trillion by 2030 every year if they're really going to make a serious dent in the SDGs, in climate, and in nature. So, how does this $5.4 trillion break down?

I just want people to be clear about these numbers. That $5.4 trillion is broken down between other SDGs and climate and nature-related investments. And I want to be quite clear in saying that making these kinds of divisions is somewhat artificial because there is so much overlap between what one calls climate-related investments and what one calls SDG-related investments. Nevertheless, just for clarity, I don't want people to forget how important it is to maintain the investments in other SDGs. And all of these conversations that we've been having about increasing skills, etc., all go to this notion that education, health, and skilled labor forces are absolutely essential for these new economies.

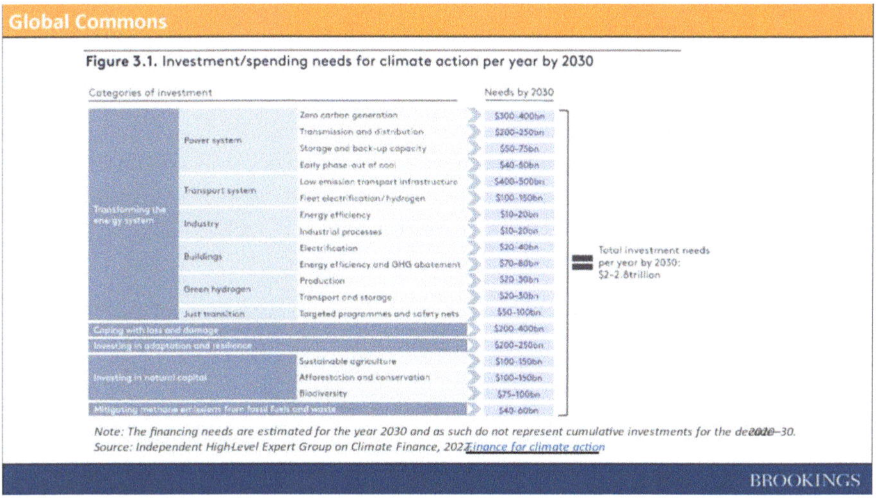

But the big increase, because we've started from such a low base, is actually in the climate and nature-related investments. And those have to be ramped up very, very rapidly. You can see from this slide that the big increase there is on the energy transition, but also with important aspects of adaptation, resilience, loss, and damage, and some of the natural capital investments that we've talked about. This is just to give you a sense of the scale of these kinds of spending. And this is probably something which is around 10% of the developing country GDP by 2030. So, it will have big macroeconomic implications if indeed it is implemented.

One can say these numbers are just plucked out of thin air. But these are numbers that have actually come from very serious, what I would call, engineering approaches, where you look sector by sector at very specific kinds of things. How much will it cost in the power system for zero carbon generation, for storage in the transport system, for EVs, for energy efficiency in buildings, and so on and so forth? So, it's built from a set of quite detailed estimates of what it would take in order to do these kinds of investments efficiently. And I was very struck actually when Kapil was talking, and he said that there are short-term upfront costs, but benefits over the long term. To me, that's just a description of what we mean by investment. Every investment involves a short-term upfront cost and benefits that come later.

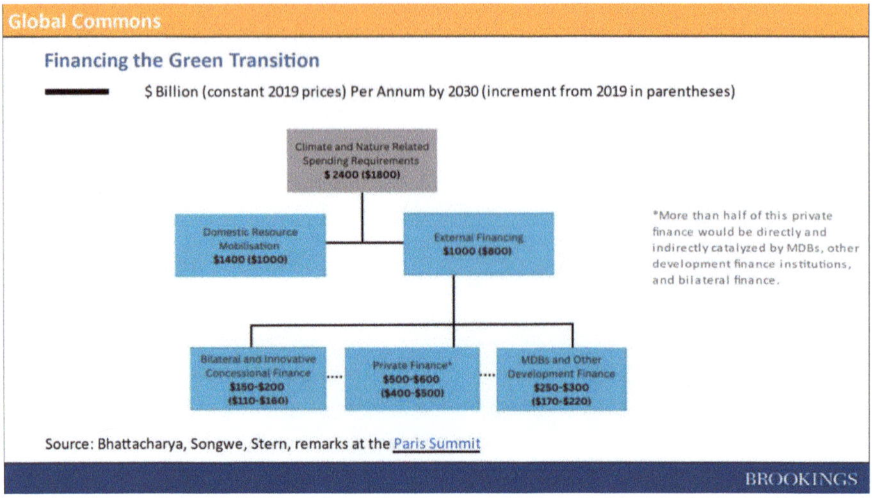

In all of these areas, we're really talking about investments. This is the nature of what we need to do. Then you have to ask yourself the question: how are we actually going to finance all of this? And so, you can start from saying this is the investment financing that is needed: some of it will come from domestic resources, some of it will come from external resources.

This is the split that we have taken in the MDB report and in other areas. And you'll see that from the external financing side, we're talking about a trillion dollars. This is the source of the famous 'we need a trillion dollars in external financing for developing countries.' What do we mean by that? We mean it's going to come from some private finance, maybe something around half. Some from the multilateral development banks and other sources of non-concessional financing, maybe about 30%. And some from concessional financing from bilateral agencies, which is also going to be quite important.

So, that's the breakdown of what we need. But I do want to underline that the notion of the Global Commons has provided a sense of scale and urgency to the question of development that we have never had before. And so, at least in the way I look at it, development is now an imperative. Before, development was always treated by the rest of the world as something that was quite nice to have. If you can manage to get some more prosperity amongst more people, that would be a good thing. But if it doesn't happen, I'm sorry, it'll happen sometime later. You can't take that same approach with climate or with nature because of the tipping points that are associated with it.

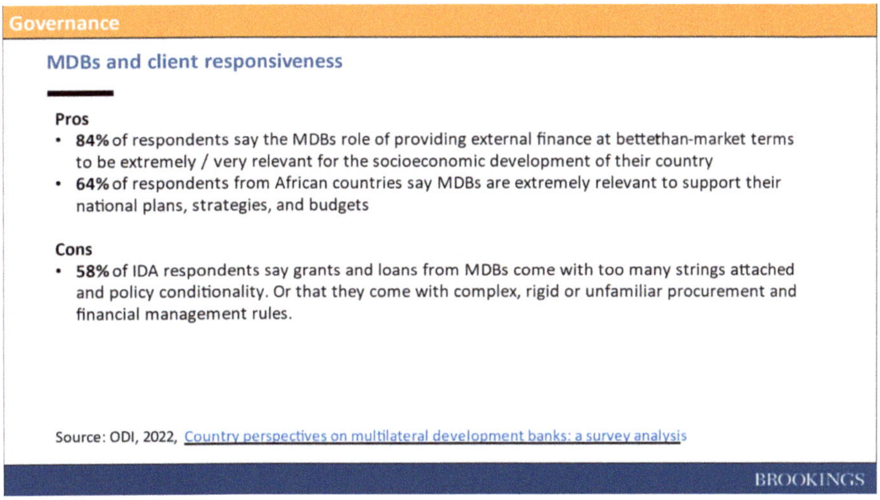

Alright, let me now turn a little bit to governance. I'm not going to talk about things like shares in multilateral organizations and their seats and stuff like that. What I'm going to talk about is governance outcomes. Are these organizations, as they're currently constituted, doing the right things, and are they doing them well?

First, I would say yes, by and large, they are doing the right things. There's a survey that was done by ODI, in which they ask, 'What are the kinds of things that you think the MDBs do which support you?' And the respondents all say, 'Well, providing external finance at better than market terms, that's very important.' Many of them say the knowledge and the technical assistance is very relevant to support their national plans, strategies, and budgets. But when they are asked, 'How are the MDBs doing?' most of them say it's very complicated and difficult to work with them.

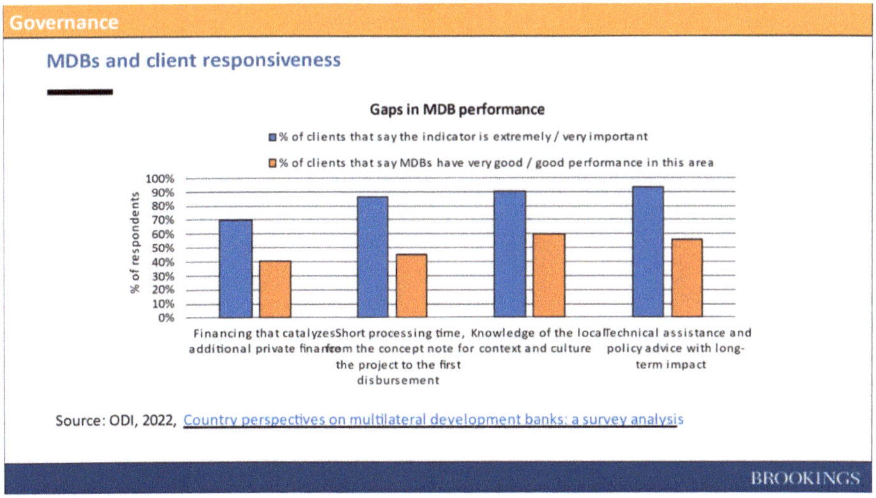

In some of the discussion I heard someone who said that he had tried to write some grants for the Green Climate Fund and what a complicated process that was. That's a theme that comes out all the time. You can see a summary of the findings of the survey. These blue bars tell you what respondents to this survey say are really important things that they think the MDBs do. And then the orange bars ask, 'Are they doing a good job?' And you can see that there are big gaps in several major areas.

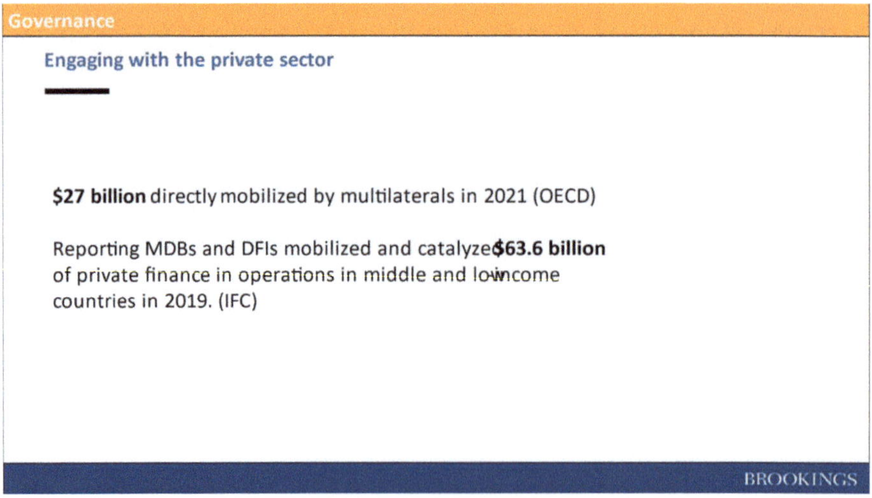

That's why we can't just say, 'Let the MDBs do more.' The MDBs also need, in my mind, a different type of governance that pushes them to improve their performance in all of these areas. One simple area where they can improve is just by being a little more transparent and clear about what they're doing. If you go into the OECD website and you ask how much private sector money is being mobilized by multilaterals in 2021, you get an answer: $27 billion. When you look at the official report by the MDBs about how much they have mobilized, they say it's $63 billion. Well, it can't be both at the same time. I understand they use different methodologies in these different places, but when we have a discussion about something so important like private mobilization, it's enormously confusing when important global bodies use radically different definitions.

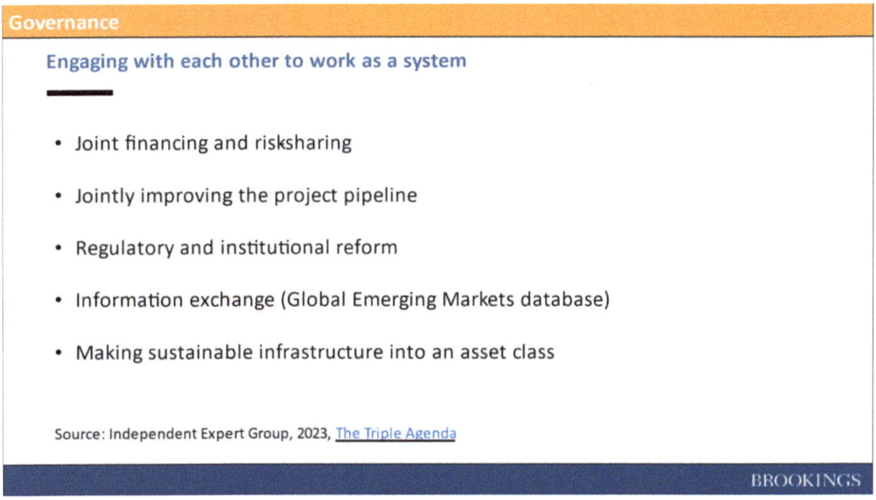

There are also lots of things, some of which we've advanced in the expert group panel report, about how MDBs could actually work much better as a system. Many people know that MDBs actually compete with each other rather than cooperate, and they do that because they're keen to finance a particular project. Right now, there aren't that many projects which are 'bankable.' Why? Because the development of the project pipeline is a public good. If an MDB develops a project and then another MDB finances it, it's a net loss for that MDB. So, they're very careful about which projects they're going to help prepare. It's only the ones they're going to finance. We've had, therefore, real underinvestment in the development of the project pipeline.

MDBs could do a lot more in terms of cooperating with each other on regulatory and institutional reforms and really pushing that agenda. They even have a database, which is called the Global Emerging Markets Database, which is about how their long-term infrastructure projects have performed in developing countries. But by and large, they keep it secret. Although my understanding is that now there's an agreement that will make this publicly available. But you need that kind of transparency if you are ever going to have sustainable infrastructure becoming an asset class. This would make the process much more efficient and would help private investors have more understanding and clarity when they get into these areas.

Now, let me just turn quickly to the politics, or if you will, the geopolitics. Where we were at the time of the last World Bank capital increase was the G7 countries were basically saying we don't want these multilateral development banks to do anything more than what they're currently doing. In fact, David Malpass—the former president of the World Bank, at that time the Under Secretary of the US Treasury— in his testimony to Congress, very proudly said, 'Yes, please give the World Bank, IBRD and IFC, a little bit more money. It's quite a small capital increase because we have got them to agree on a new financial discipline mechanism that will stop them from ever coming back to you again for more money.' So, the last capital increase really served to limit what IBRD calls the sustainable lending level to $27 billion per year in new commitments. That's now gone up a little bit, thanks to the implementation of the recommendations of the so-called capital adequacy framework, but still not by much. But the mental frame was we don't really want or need these institutions to expand enormously.

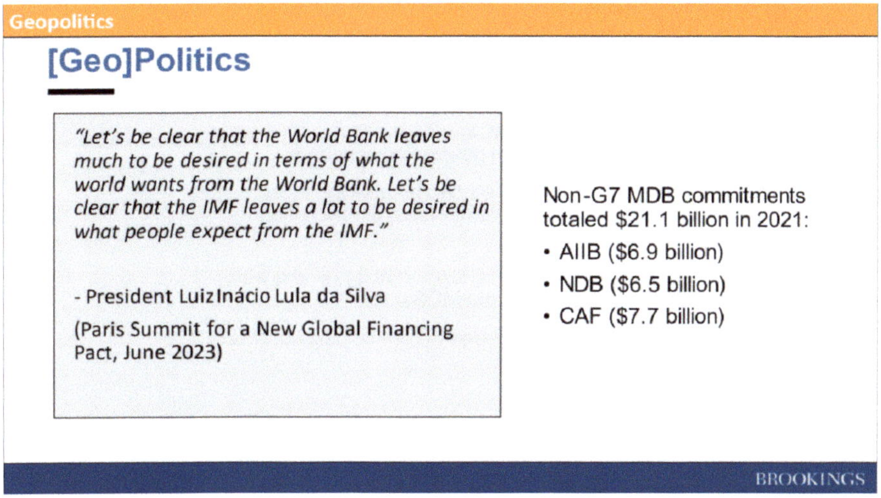

Then, on the other side, you have the developing countries—the big developing country borrowers. President Lula, when he went to the Paris Summit on the new Global Financing Pact, said, 'Let's be clear, the World Bank leaves much to be desired in terms of what the world wants from the World Bank and the IMF.' I was smiling when I heard in the last session that maybe the G20 wants to take a look at the IMF as well as at the multilateral development banks. Certainly, that's something many developing countries want.

But what was the response? The response from the big borrowers was not to say, 'Let's try to fix these institutions.' It was to say, 'Let's create our own institutions.' And the big new institutions—the Asian Infrastructure Investment Bank, the New Development Bank, and CAF (what's now called the Latin America Development Bank)—which don't have any G7 shareholders (though AIIB does), were founded by developing countries for developing countries, and now are committing something like $20 billion a year. So, quite a large portion of the total MDB commitment.

My point here is simply to say there is not necessarily a huge amount of political support for the current multilateral development banks in the environment we see ourselves in today. This is rather unfortunate, and I'm going to conclude with this because, as I see it, here are a few propositions which I think many people would agree with. The first is you cannot meet these global challenges without a very sizable step up in public and private investments in developing countries. I don't think many people would disagree with that statement. Then you say, 'Could developing countries just do this by themselves?' And again, with very few exceptions, maybe China might be able to do it by itself, but with very few exceptions, because of their limited financial and technical capacity, I would say almost all will need considerable international support.

Then you ask, 'Where is this going to come from?' Some people have talked about new institutions. It takes years for a new institution to become operational and to be able to scale up its lending. We don't have years to really start addressing these global challenges at scale. You can't do it just through concessional lending, the Green Climate Fund, etc., because there isn't enough money there. They don't have the leverage. So, I believe that the only option we have to deliver the support at scale is the multilateral development banks because of their ability to leverage financial support from their shareholders. But we won't. Nobody is willing to put more money into these shareholders unless the MDBs, in turn, transform themselves. And they have to do that before helping countries to transform the world.

Reference

1. Independent Expert Group of the G20 "Strengthening Multilateral Development Banks; the Triple Agenda", Volume 1, July 2023 https://www.g20.org/content/dam/gtwenty/gtwenty_new/document/Strengthening-MDBs-The-Triple-Agenda_G20-IEG-Report-Volume.pdf

Chapter 6
Multilateralism: Geopolitics, Governance, and the Global Commons

V. Anantha Nageswaran, Jean-Louis Arcand, Mari Pangestu, Ram Madhav, and Ramesh Chand

6.1 Session Chair: V. Anantha Nageswaran

We all know it's a truism that there is no politics without economics, and there is no economics without politics, and there is no multilateralism without taking geopolitics into consideration. We know that multilateralism flourished after the end of World War II. There were attempts to create multilateral institutions in the aftermath of World War I as well, but that met with a little success. In fact, if we analyse the causes for multilateralism to take off and flourish after World War II but not after World War I, we might in fact be able to understand some of the reasons as to why it is currently floundering.

In fact, at the end of World War I, the kind of obligations and reparations imposed on Germany created the preconditions for the emergence of World War II indeed. So, are we in one such situation now? Because post World War II, the victors, although the camp split immediately after the end of War II, the victors were able to take an attitude of enlightened self-interest and combined with magnanimity, were able to create several institutions that stood the test of time for several decades. They have currently commanded strain. In fact, we are having this discussion within a week or

V. A. Nageswaran (✉)
Chief Economic Adviser, Ministry of Finance, Government of India, New Delhi, India

J.-L. Arcand
President, Global Development Network (GDN), Geneva, Switzerland

M. Pangestu
Former Managing Director, Development Policy and Partnership, World Bank, Jakarta, Indonesia

R. Madhav
President, India Foundation, New Delhi, India

R. Chand
Member, NITI Aayog, Delhi, India

S. Bery et al. (eds.), *Navigating Challenges for Sustainable Growth*,
https://doi.org/10.1007/978-981-97-7894-2_6

two of two important books having been released. One by Neil Howe, who along with

William Strauss co-wrote The Fourth Turning, and another book by Peter Terrence, both of which have been reviewed by Francis Fukuyama for New York Times.

So, is this an inevitable part of the cycle that nations and the globe itself goes through? In other words, do things have to get worse before they get better? Alternatively, in order to make sure that multilateralism works in the interest of global commons and global public goods, for example, do we need a Truth and Reconciliation Commission for Global Commons similar to what South Africa did at its independence on topics such as global warming and pandemic prevention?

Therefore, should be the topic of this session. How has geopolitics affected multilateralism and the governance of multilateral institutions? How far and how much has it come in the way of the provision of global commons and global public goods?

6.2 Speaker 1: Jean-Louis Arcand

I'm going to talk in the context of this session about health and vulnerability, something which has been patently absent from what we've talked about over the past day. Perhaps in some sense it's because we're all tired of dealing with health issues after the Covid pandemic. It's like our British friends with Brexit; they just don't want to talk about it anymore. But at some point, especially in the context of the G20, we do have to talk about health vulnerability, what the G20 can do about this, and the link that it has to the Global South.

This presentation will have a three-pronged structure. First, when health involves positive externalities, multilateralism, and the global system seem to have no trouble handling it. These are no-brainer gains. I'll give three examples that I've worked on, but there are many more, and a lot of people in the room could come up with other examples.

Second, I'll discuss vulnerability to losses, taking the example of Covid, but not in medical terms or vaccine nationalism. Rather, I'll focus on the economic costs of Covid to the Global South. This will link up to what Murray started to say before me. We have a lot of good intentions, especially from the Global North with respect to the global south, and maybe there's something we can do about this.

I'll conclude by trying to be a little provocative in terms of the G20 and geopolitics. The conceptual framework that economists use to think about health will be the basis for my arguments.

Health and development are often understood through what's known as the Grossman model, where health serves dual roles. It's both a consumption good and an investment. On one hand, we value health because it makes people more productive. On the other hand, it's a consumption good in its own right that contributes to well-being.

When we consider health interventions, we're not solely interested in their economic impact. Examples of such interventions include Universal Health Coverage

(UHC), which is a major topic on the global stage. Having spent the past 15 years in Geneva, I've had numerous interactions with WHO experts who frequently discuss UHC. Malaria eradication and the introduction of various types of vaccines, such as a novel TB vaccine set for introduction in 2028, are other examples.

These interventions often have favorable cost–benefit ratios. While their impact on economic growth may sometimes be relatively small, their effects on health metrics such as Disability-Adjusted Life Years (DALYs), Quality-Adjusted Life Years (QALYs), or lives saved, and deaths averted are significant. Multilateral institutions are well-equipped to address these issues, given the high payoffs of these interventions.

We have global institutions like the WHO and more specialized ones like GAVI or the Global Fund that coordinate action on these global health issues. For example, let's consider Universal Health Coverage (UHC), which targets a range of health issues from childhood diseases to non-communicable diseases, TB, and HIV/AIDS. I've worked on this with colleagues at the WHO, and when you look at the impact of UHC on GDP or sustainable growth, it's not necessarily huge. However, the impact on averted deaths can be significant. For instance, implementing UHC in India could prevent a large number of deaths, as illustrated in Table 6.1.

Another example is malaria eradication. There's a whole body of literature on this, and global eradication is indeed possible. We've made significant progress in this area. Eradicating malaria not only saves lives but also has economic benefits. It shifts the income distribution of endemic countries to the right, effectively improving the economic conditions of these countries. This is illustrated in Fig. 6.1.

So, these are what I would call "no-brainer" investments. Depending on how you value human life and well-being, there's a strong global will to implement these kinds of health interventions.

The novel tuberculosis vaccine is another example of a global good where multilateral systems can help us adopt these interventions. Especially in Southeast Asia, India, and Russia, the introduction of this vaccine is expected to yield significant

Table 6.1 Averted deaths due to UHC

Year	2019	2020	2021	2022	2023
India					
Baseline Ambitious	10′356′891 9′743′562	10′658′766 9′886′636	10′932′837 10′022′703	11′192′963 10′157′449	11′441′065 10′304′529
Averted deaths %	613′329 5.92%	772′130 7.24%	910′134 8.32%	1′035′514 9.25%	1′136′536 9.93%
China					
Baseline Ambitious	9′402′246 9′351′203	9′799′029 9′732′516	10′168′493 10′088′984	10′514′332 10′425′975	10′843′400 10′749′849
Averted deaths %	51′043 0.54%	66′513 0.68%	79′509 0.78%	88′357 0.84%	93′551 0.86%

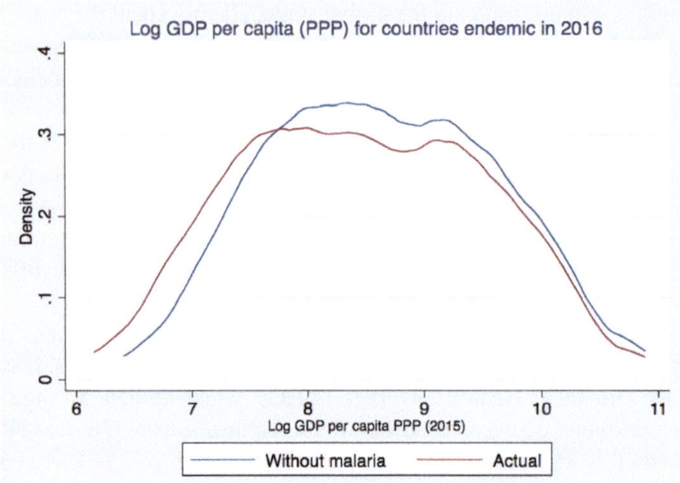

Fig. 6.1 The economic gains from eradicating malaria

benefits. Again, these are "no-brainer" investments that the global community should be making, as illustrated in Fig. 6.2.

Now, let's switch gears and talk about the flip side, which is Covid. The issue here is that we have permanent impacts from transitory shocks. I'm not talking about the vaccine itself, but rather the economic consequences of the pandemic. In the global North, including OECD countries and China, we had shutdowns for very good reasons. These shutdowns led to massive recessions, with an approximately eight percentage point drop in GDP growth.

The shutdowns were informed by epidemiological models, and many of us in the economics profession suddenly found ourselves diving into epidemiology to understand the dynamics of the pandemic. But the key point is that these shutdowns, while necessary for public health, had severe economic repercussions.

What happened in the Global South, particularly in sub-Saharan Africa (SSA), is that the recession in the Global North was transmitted to these regions. Estimates

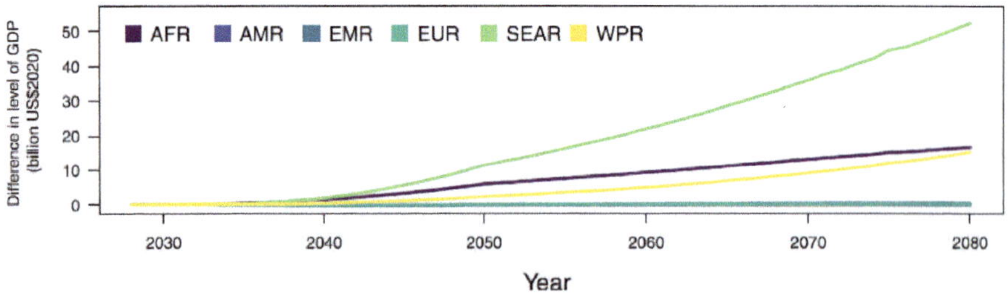

Fig. 6.2 The expected economic gains to introducing the novel tuberculosis vaccine, by world region

suggest that about four percentage points of growth were lost in sub-Saharan Africa due to the recessions caused by lockdowns in the global North. The primary transmission mechanism was trade, especially affecting countries in sub-Saharan Africa that were more open to trade.

Figure 6.3, based on IMF World Economic Outlook data, shows the predicted growth for Africa in November 2019 and then what was expected in April 2021, after the onset of Covid. The impact is clearly visible.

The third part of this mechanism is that severe recessions in the Global South, and particularly in sub-Saharan Africa, have fatal consequences. These recessions are not just economic statistics; they have a human cost.

The data show that severe recessions have a direct impact on mortality rates, particularly in sub-Saharan Africa. While in developed countries, health tends to be countercyclical with respect to GDP, the opposite is true in developing countries. The graphs in Fig. 6.4 indicate that severe recessions, even when controlling for GDP per capita levels, lead to increased death rates and infant mortality. Positive

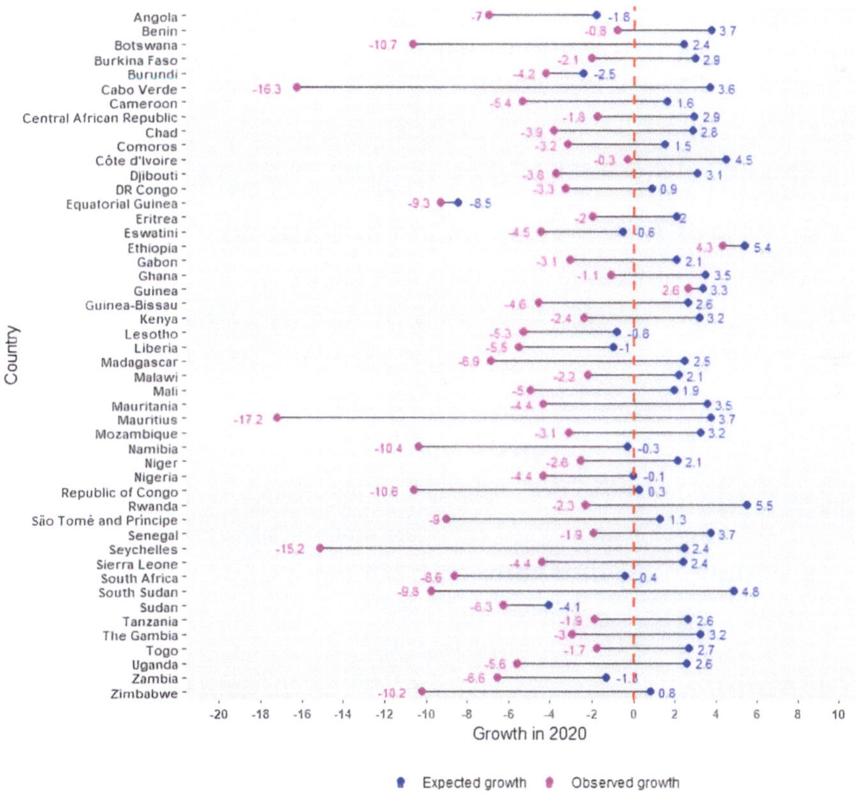

Fig. 6.3 The growth shortfall in SSA during the onset of the COVID-19 pandemic

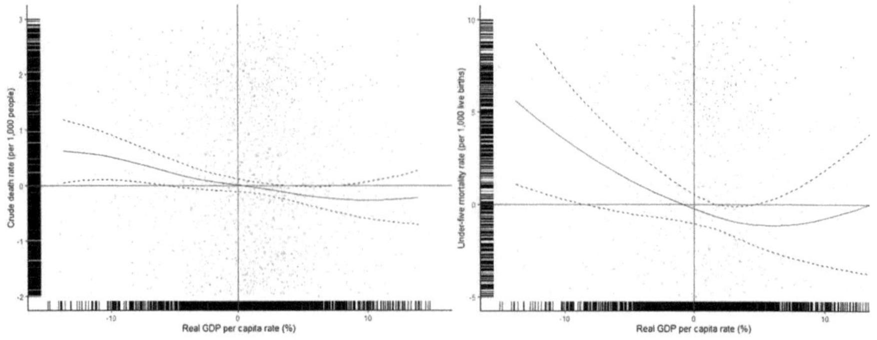

Fig. 6.4 In SSA, severe recessions kill, but expansions do not save lives

growth, on the other hand, doesn't significantly reduce mortality, as indicated by the flat confidence spans to the right of the graphs.

When you put all these numbers together, the back-of-the-envelope calculation suggests that the actual cost of Covid to just sub-Saharan Africa, due to shutdowns in the Global North, is roughly half a million lives lost. This is illustrated in Box 1 which provides details of the calculations. This is a significant transfer of human costs (death, to be precise) from the Global North to the Global South.

Two sobering back-of-the-envelope calculations

- **95% of Africa's population is under 60**
  Fall in African growth due to lockdowns in the OECD $\times$ marginal effect of African growth on under 60 mortality rate $=$ increase in under 60 mortality rate
  $$4.14 \times 7.87 = 32.58$$
  $\Rightarrow \frac{32.58}{100{,}000} \times 1.27 \times 10^9 \approx 414{,}000$ additional African deaths due **solely** to lockdowns in OECD countries (Compare! 100,000 African deaths attributed to COVID)
- **Infant (0-1) mortality rate** (71 per 1,000 in 2020)
  Has been increased by: $4.14 \times 1.408 = 5.83$
  Lockdowns in the OECD have essentially wiped out the gains of the past 5 years...

Box 1. The death transfer from the Global North to SSA. Marginal effects of African growth on African mortality come from the author's econometric estimates, which are available upon request.

I'm not arguing against the necessity of shutdowns in the Global North; that's not the point. The point is that these spillover effects are not adequately considered by the international community. There are numerous examples where policies in the Global North have unintended, often devastating, consequences in the Global South. This highlights the need for some kind of international insurance mechanism to mitigate these impacts.

Points that have been raised earlier, such as the call for a more focused and effective approach from the G20, are well taken. The decline in Official Development Assistance (ODA) over recent years is of great concern, particularly for sub-Saharan Africa. While ODA may not be a growth engine, it can serve as a crucial insurance mechanism, especially if designed to be countercyclical. This could help mitigate the negative spillover effects of policies in the Global North on countries in the Global South.

I have provocatively argued elsewhere for the equivalent of a "Non-Proliferation Treaty" on Sustainable Development Goals (SDGs). The current framework with its 169 targets and 247 indicators is overwhelming and, frankly, impractical for policy implementation. A smaller group like the G20 could indeed focus on more tangible goals, such as sustainable growth, which would be more actionable and impactful.

Adding an insurance mechanism specifically for the world's poorest countries might be an idea worth delving more deeply into. Whether we're talking about sub-Saharan Africa or the Least Developed Countries (LDCs), such a mechanism could provide a safety net against the unintended consequences of global policies. This could be a significant step toward a more equitable and sustainable global system.

6.3 Speaker 2: Mari Pangestu

I'd like to make three points in my remarks.

1. First, why multilateralism and maintaining the global economic order is still key, given all the challenges we have.
2. Second, what is the kind of multilateralism and global economic order we need to have?
3. And finally, what it means for the institutions and regional multilateral institutions and processes, including regional institutions and processes that we have now.

So, let me start with why multilateralism and maintaining the global economic order is important. It's very clear from all that we discussed yesterday that the global economic order is at risk of becoming frayed and dysfunctional because of many reasons. One is the decline in US leadership and the shift in hegemonic interest and domestic constituents. Populism's rise was mentioned yesterday, which makes the US less willing or capable of leading the public good outcome of the previous global economic order of an open and rules-based economic order.

You have the China-US competition and geopolitical tensions leading to new divisions. Yesterday, we had several presentations showing the structural changes in the world. We are moving from a unipolar to a multipolar world of economic power with a clear emergence of developing economies. This is why the discussion about the global south in the G20 year of India becomes very relevant. And given this threat and the global challenges we face, we need multilateralism more than ever.

Multilateralism, I would argue, is the prerequisite to managing the transition from a unipolar world to a stable multipolar new order, not a US block of friends or allies, not a China block, and neither a global south block. It has to be a block pursuing agreed, shared principles and rules. We still need rules and norms to stop free riding and bad behavior. The whole global economic order and rules-based system were intended to stop or constrain bad behavior by members and the leader. Now, we are seeing the rise of protectionism and WTO rules like non-discrimination being violated, leading to beggar-thy-neighbor policies. Yesterday, we had many comments on the US industrial policy and how it breaks the WTO rules, leading to provoking action by others.

So, multilateralism is also needed to address the global common challenges, whether it's climate, health, or the slow growth trajectory that's hurting developing countries. I would argue that trade and investment are still crucial parts of that growth story.

Secondly, what kind of global economic order do we want? We want one that continues to deliver not just growth but sustainable, resilient, and inclusive development. There's wide agreement on that. And peace, or as it was originally termed in global public goods, was growth and peace. But it's about sustainable development. It's about being resilient enough to manage global risks and disruptive technological innovations. We heard a lot about that yesterday: climate, technology, and I would add the debt crisis, energy, and food security. And finally, it should be cohesive enough to manage systemic crises. We learned a lot about that during the pandemic and now as we face the climate crisis.

Thirdly, therefore what kind of multilateralism and multilateral institutions are going to be fit for purpose given the current context. We have to take the current situation as a given, which is the intermix between security concerns, geopolitics, and the ongoing US-China geopolitics that aren't going away anytime soon, and the social and political conditions facing countries. This is the starting point and the answer is a combination of maintaining multilateralism, open regionalism, and international openness, which are essential to secure sustainable development and peace. This is the main outcomes we would like to achieve.

Now, let's have a discussion about what that means moving forward. What does it mean for the governance of different multilateral and regional institutions? I think multilateralism and international cooperation to address shared global challenges relative to traditional and non-traditional security challenges are going to be important. There's an opportunity to cooperate based on shared interests. I'm an optimist. Many are very pessimistic about multilateralism, but I'm a diehard multilateralist and probably an optimistic diehard multilateralist. I look at the glass as half full. I believe that while many things didn't work in the multilateral and global economic order, it's not about finding a new order or overhauling the institutions. It's about building on, strengthening, and evolving the existing institutions and processes.

It is also about international cooperation and support to ensure concerted unilateralism—or how countries act nationally. If you look at what's happened in the last eight decades, countries do sign on to the WTO agreement, but what happens is countries are the ones that take action. Countries signed up to Paris Climate Agreement, but again action happens at the country level. Some agreements are binding, like the WTO with clear commitments and sanctions for not fulfilling as well as at least in the recent past clear dispute settlement procedures. Some are not binding such as the Paris Agreement. To achieve unilateral and converted actions, international cooperation, international discussion and dialogue, and the conversation to agree on what we need to do and how we need to get there are key. Confidence to act comes from others doing it. Confidence comes from capacity building. Confidence comes from not just the promise and pledges of resources but real resources and support coming through if you did what you agreed to do.

The other dimension, which we didn't really talk about but touched on it a little bit yesterday, is that in the new multilateralism is not just state to state, but has to include the the private sector. The technology story was very clearly a private sector issue, and self-governance also in the private sector and how to incorporate society.

How should we then transition to a multipolar world, without US leadership, and accounting for geopolitics? It should be about the shared interests of the global south and the middle powers who should not want to be bifurcated by the US-China conflict. Their shared interest is still about open trade and investment and about navigating and managing the bifurcation between the US and China. Yesterday, there were powerful presentations about the costs of decoupling. If you had total decoupling, the cost is very high. And guess who pays? It's the developing countries that will pay for what emerges. Whether it is decoupling or de-risking, economic instruments are being used to address security and dependence on China concerns. Whilst recognizing the valid security concerns, how should the policies and instruments be managed that can minimize the distortions and damage that it can cause?

The first best is not to have all the intermix between economics, security, and technology, but let's face it, the reality is there and there are valid concerns. So, second best, how do we minimize the damage?

There other important shared interest of middle powers and the global south is the green transition. Yesterday's discussion was a powerful endorsement for the green transition coming from developing countries. That's the big shift in the last few years where developing countries were saying no, that climate is a luxury and we don't have the resources to focus on climate, so we need to focus on growth. Now the whole debate has totally changed. Having been very involved in this in the last three years, we do see developing countries recognizing that climate and development are not tradeoffs, that one can achieve development whilst addressing climate and in fact not addressing climate will come at the cost of development. But developing countries are asking, "Okay, we now see that we need to address climate, otherwise we will not develop, but we need resources, we need technology, and so on."

Given these shared interests, how should the international institutions and governance be reformed, evolve, and strengthened and continue to agree on existing and renewed principles, norms, and standards. It will have to be a multi-pronged approach building on various existing systems and processes, not just one or the other. Let me give you the example in international trade, because that's my area of knowledge and expertise.

The resilience of global value chains, facilitating investment, unlocking ICT, and pushing back on protectionism, WTO reforms, including dispute resolution, were key in the G20 agenda. India would be hurt in its green transition if it's not integrated with the world economy. I'm quoting from the statements of Indian officials from yesterday.

So what does it mean for the processes and the institutions that we have? I'll start with the G20. The G20, hopefully, can overcome the challenges. Since 2016, the G20 has been somewhat absorbed in communique language. During the Trump years, it was about removing everything that said "multilateral." Now we know paragraphs three and four are about war. It's unfortunate, but hopefully, we can rise above that and delve into the substance. The G20 is about agreeing and demonstrating the political will of what needs to be done collectively, what are the principles, norms, and standards. And then it's taken elsewhere to negotiate, whether it's the WTO or other fora, or the G20 acts on it, like the concerted fiscal stimulus in response to the global financial crisis. So I do think the G20, especially taking into account the concerns of developing countries, continues to be key.

How do you give the global south a voice? Let me share three thoughts. The global south can be at the table, and some of them are at the table in the G20, or they could be represented at the table. For instance, Indonesia represents ASEAN, and we always get ASEAN to be invited as an observer, even if you're not at the table. One can think of other groups that can be represented in a similar way. These are just some ideas, other than G21.

Let me now focus on trade briefly and what a multi-pronged strategy would look like. First multilaterally, the focus is on WTO reforms and this has been on the G20 agenda since the Osaka G20. One priority is to address the dispute settlement and appellate body issue, because this is an important issue for the global south—where small and big countries have the same right to a fair, open, and rules-based trading system. In terms of addressing other issues, in parallel plurilateral initiatives or what we call "Club of Clubs" initiatives should also be continued to be galvanized. One example is what's called MPIA, multi-party interim appeal arbitration, which involves 53 countries, including China, Japan, the EU, and Australia. It's essentially mirroring the appeal process of the dispute settlement process, but not within the WTO. So, it's about keeping the bicycle moving in terms of promoting open, rules-based, inclusive trade. Another area that I mentioned yesterday is about the use of the security clause under GATT Article XXI. How can we codify the grounds for using security in restricting trade? This will likely spill over into defining the criticality of goods for which you can impose trade restrictions or temporary bans. Finally, outside of the WTO, the "club of clubs" can be thematic, like what's happening with services, e-commerce, and digital trade. Even if the US isn't participating, progress can still be made.

I would also strongly advocate for regional trade agreements. As mentioned yesterday, deep regional trade agreements can fill the gaps left by the multilateral system. These agreements can encompass competition policy, especially given the rise of large technology companies, as well as services, labor, and environmental standards. Trade and climate is another dimension that can be addressed both multilaterally and regionally. Many Asian agreements emphasize partnerships, such as the Regional Comprehensive Partnership Agreements, because they focus on capacity building and accommodating varying levels of development. True capacity building, combined with financial and technological support, is essential.

My final call to action is to revisit a paragraph we discussed a couple of years ago. It essentially states that G20 countries support a rules-based multilateral trading system but also endorse regional or alternative pathways to achieve this in an open and inclusive manner. It's about open regionalism, ensuring that even plurilateral agreements are inclusive, allowing observers who haven't signed up to participate in discussions.

6.4 Speaker 3: Ram Madhav

As the leader of the G20 Group of Nations, in this ongoing year, India tried to share its wisdom with the rest of the world. So, let me begin my presentation also with one of India's ancient beliefs about what we today call as global commons. Our ancient wisdom always stated that the creation was made of five elements. We called them in Sanskrit as Prithvi/Bhudevi (Sanskrit: पृथ्वी, Earth), Apas/Varuna/Jala (Sanskrit: आपः, Water), Agni (Sanskrit: अग्नि, Fire), Vayu (Sanskrit: वायुः, Air), Akasha/Dyaus (Sanskrit: आकाश, Space/Atmosphere/Ether/Sky). It meant our earth, our oceans, our energy sources, our environment, and our space and outer space and cyberspace. We believed that these five elements constitute the entire creation, and hence, they should be out of bounds for sovereign national governance. This has been the ancient thinking of this country. What we are calling as global commons today have always been held for the benefit of the entire humanity.

So, when we created the global multilateral institutions, the most successful ones after the Second World War, in the form of the United Nations and its many allied organisations, the original focus should have been about these five elements. It has been, I'm not denying that, but the effort has not been so very successful for various reasons. One reason one can attribute to is probably the bipolar Cold War politics in the initial three to four decades. Subsequently, in the last two to three decades, what we see is the general perception, that the multilateral institutions that we created are not able to deliver what is expected on issues like global commons. In fact, on many issues. But since our focus is on global commons and that they're unable to deliver fully. Hence, India also maintained this position that probably the time has come for us to now think about a complete overhaul or restructuring of these institutions.

One important reason, I believe, or an argument in favour of restructuring of these multilateral institutions that we created, is because from a bipolar polity, we moved to a multipolar setup today. Today, I would probably go one step further to say that we are not even a multipolar world. We are a heteropolar world. Today, multipolarity as a reality needs to be acknowledged because there is no one or two countries that would finally determine the destiny of all the countries of the world. There are many powerful countries emerging in different parts of the world. Many mini-laterals have emerged. We are a part of SCO, Shanghai Cooperation Organisation. We are a part of BRICS. There is ASEAN. There are other groupings. These mini-laterals also have emerged as important poles in the world.

Besides this multipolarity and emergence of mini-laterals, we today see the rise of big tech, rise of multinational economic corporations, rise of global NGOs, and of course, there are organisations, there are terror groups, there are religious organisations. All of them defy national boundaries. For them, national sovereignty means nothing.

Our internet is no longer easy to control. You have to be a China to control the internet. So, we are living in a totally different kind of world order, a heteropolar world order. This has to be acknowledged as a reality today.

I see some resistance to this idea. No, we are still led by so and so. So here, the question of how do you, when it's a multipolar world, attend to the aspirations of all the different countries in the world? Is it up to the United Nations to be able to do that, or do we need to have a totally different approach to this whole issue of global governance? That is an important issue to take up for discussion. That is where India's championing of the cause of the Global South becomes important. This Global South, as we all know, consists of countries which are developing in nature. We used to call them the developing world. We are calling it the Global South now, but 80% of the world's population, that means the so-called global commons, will be positively and negatively affected by these countries.

Majority of these countries, for example, countries in Africa, are Small Island Developing States (SIDS). They depend on what we call the global commons for their survival. You cannot deny them access to them. So, the challenge is, how do you lead this whole Global South and its aspirations to grow?

Here, India wants to take the lead. Essentially, the Global South and the mini-laterals that have emerged in many parts of the world have to be accommodated in any new structure that we build for future global governance. Any future United Nations should not be just a body of 195 or 197 countries alone. It should be able to represent the aspirations of this heteropolar world order. So that is where probably India's leadership of the Global South will lead us.

I will just leave three issues before you for consideration when we try to rebuild the multipolar order, the heteropolar order for our future multilateralism. Number one, as I said, how can we make these heteropolar components a part of our future multilateralism? Second, the Global South is a developing world. As I said, it depends on the so-called global commons, especially the oceans, the energy sources, and the earth for its survival. Now, suddenly we cannot say, no, you cannot access them, or suddenly you have to reduce emissions. Of course, they all have to come onboard and do their bit, but is it at the cost of their aspirations to develop? Do we want those countries to remain underdeveloped, not developing? Because I know about India, for the last three decades, we have been called only as a developing country. It has been almost three decades now. Our Prime Minister wants India to be called a developed country 20 years from now.

But that aspiration of the countries to become developed countries requires dependence on the very same resources which we're calling as global commons. How do we compensate if we want them to reduce their dependence on these commons? How do we compensate? Are we willing? Are those who have developed, who have progressed, willing to shell out some pounds, some dollars for that purpose?

Last, but I consider it as the most important, is: how do we bring social sciences and natural sciences together? When I say social sciences, I mean our economics, our education, our technology, our politics; they take a particular direction. But natural sciences have a different focus. How do we bring these two together? Every economic activity that we undertake from this point, how is it going to be aligned with our nature, our environment, our global commons? In the age of AI, how do we marry these two things? This is going to be a big question before all of us in the next 20 to 30 years, I believe.

So, friends, I will end by saying that the need for a new, renewed, or revised multilateralism has been emphasised by all of us. But that new multilateralism cannot be just that of a few countries of the world led by someone. It has to take into account the current hetero reality and aspirations of the Global South.

6.5 Expert Comment: Ramesh Chand

If we want to discuss or debate the future of multilateralism, we need to delve into developments that happened in the post- WTO period. Two parallel movements followed the post-WTO period. One was the expansion of multilateralism and second was simultaneous explosive growth in PTAs—you call them free trade agreements, regional trading agreements, but I find them neither free nor regional. Therefore, I'm using the term preferential trading agreements. Up to 2007, when the global financial crisis happened, these two movements just went in the same direction. But after 2008, the multilateralism started weakening, fragmenting, but excitement among nations about preferential trading agreements remained strong, and their growth continues.

I feel based on this experience, we need to clearly draw out what the core areas for multilateralism are, which cannot be dealt with by PTAs. We know it: if we want to address the effect of subsidies in fisheries, we know it cannot be addressed through a PTA or a regional trading agreement. We have to turn to multilateralism.

I feel it is very important to understand whether the growth of PTAs has in some way adversely affected multilateralism. The old debate: wether PTAs or RTAs are building blocks or stumbling blocks for multilateralism. I feel we need to draw some lessons, and what are the core areas for multilateralism that will help in building a strong case in favor of multilateralism.

Then secondly, yesterday and also today some references were made toward a shift from globalisation to de-globalisation, and the slowing down and fragmentation in multilateralism. I feel that we need a little more nuance in this because when we talk of multilateralism broadly, the issues can be divided into three categories. One category includes issues related to economy, trade, banking, fiscal matters, and the like. The second is issues which are related to humanitarian aspects: health, nutrition, food security. And the third category includes issues which relate to the survival of people and planets such as climate change, sustainability, environment, biodiversity. I feel the presentation which was made yesterday on fragmentation of multilateralism, weakening of multilateralism, applies in my mind to the issues in category one.

As far as issues in category two and three are concerned, I find that the global mechanisms to address them are getting stronger. SDG, whatever it is, followed after MDGs in 2015. Even with SDG limitations, we are discussing food system transformations and there's a lot of interest in decarbonisation and how we do it.

So, I think we should not sweep it with a broad brush that multilateralism everywhere is becoming fragmented or weakened. Yes, in the area of trade and economy that is happening. But in the other areas, this trend toward multilateralism, I find, is rising. Second, diagnosis of underlying factors for deglobalisation or slowing down or fragmentation of multilateralism is very important. To what extent it was because of slowing down of growth in global economy, to what extent it is due to policy decisions taken by different countries? That distinction is very, very important.

I looked at some of the data, like if we look at trade intensity or the share of the ratio of export to GDP since World War II till 2007, it has been increasing and the peak was reached at 67% in 2007. Then the world faced global financial crisis. The trade intensity dropped suddenly from 67% to 52%, and afterward settled around 54%. We did not see any decline in global trade to GDP ratio in the subsequent 15-years period, but it kept fluctuating around 54%. So, is it that after the global financial crisis, the growth in the global economy did not recover, and that is the reason for decline in trade intensity, or, the reasons are more policy induced, like Brexit, trade restrictions, etc.? I think there is a need for strong empirical work to understand this change.

Also, I saw one IMF study which says that the effect of mild fragmentation on global economy will be a reduction in GDP by 0.2%. But in case of severe fragmentation, it can go up to 7%. But if you take into account the other channels through which the effects percolate down, then it becomes much more serious.

Another issue that I want to share is that many developing countries agreed to be members of WTO despite a lot of opposition at the domestic level. But after some time, frustration started building with the feeling that the playing field was not level. As developing countries were not able were not able to have a say in the WTO, the stalemate began and took its toll. I feel this also needs to be factored in to find out whether the stalemate in the WTO is one of the reasons for the reduced interest in multilateralism.

I will quickly make two more points. One big change, which is happening in the global economy, and from there I want to link it to my area of interest, that is agriculture. There's this structural change in the economy which was happening as per the Arthur Lewis hypothesis, that as economies grow, the share of agriculture in GDP declines. That trend stopped after 2005. Between early 1970s to 2005, the share of agriculture in the global GDP declined from 10% to 3.2%. But after 2005, the share of agriculture, rather than declining, increased by 33%, from 3.2% it went up to 4.4%. I feel this has very serious implications for development policy, for trade, even for multilateralism, even for climate change, and other such issues. I keep discussing these things with colleagues in NITI, and others, to reimagine agriculture and its role in development.

The changing nature of structural change in the output of global economy has serious implications. Even more serious is the impact on employment because manufacturing growth has slowed down, and manufacturing is going for more capital-intensive production. The labour shift from agriculture to non-agriculture has either halted or it is very slow, and this is a matter of concern.

Somebody mentioned about G20 having so many issues. I made a presentation from Indian side in the meeting of G20 Agricultural Ministers in Hyderabad, where we pointed out that the most serious issues in the AgriFood sector is that after 2015, hunger and undernutrition in the world has started increasing. The reversal in hunger and nutrition improvement started around 2010-2012, first in Africa, then it spread to South America, and of late, it has hit South Asia also. And what is more disturbing is that the hunger and undernutrition have started deteriorating despite the fact that agricultural growth is intact and per capita food production is rising!

So, these things, I think, need to be taken into consideration. They are vital. I mention these things not only for the sake of agriculture, but also for the sake of other issues like technology transfer from developed to developing countries, from north to south, affecting costs of production, and many other such factors.

I think there is another issue worth considering. Shree Ram Madhav ji was alluding to global commons, which are existing entities, like our oceans, forests, and biodiversity. These entities are here, and they come with a set of livelihood and sustainability issues related to them. That is one set of issues. However, I think in this particular forum, we are more concerned about global commons that we aim to create, like initiatives for decarbonisation and sustainability.

Chapter 7
Adjustment, Resilience, and Inclusion in an Uncertain World

François Bourguignon, Santiago Levy, Haroon Bhorat, Surjit Bhalla, and Vinod Kumar Paul

7.1 Session Chair: François Bourguignon

I believe that the evolution of global inequality over the recent past is a good illustration of the various themes we want to handle in this session. You may know that a potent change took place in global inequality of living standards over the last two decades or so. After having risen almost continuously over the last two centuries and reached a very high level, much above the most egalitarian countries in the world, a reversal took place at the turn of the millennium, and global inequality actually began to fall at a very fast speed. In a few years, it erased almost a complete century of increase in history. To some extent, some people considered this a kind of historical turn in the world.

Initially, this development was driven by the performance of big emerging countries, China in the first place, and then quickly India. But at the turn of the century, really all developing countries were involved in this process of catching up over the most advanced countries.

Again, this was a huge change taking place almost in the global order. And this seemed to be rather robust in the sense that even the Great Recession in 2008–2009

F. Bourguignon (✉)
Chair, GDN Board; Professor Emeritus, Paris School of Economics, Former Chief Economist, World Bank, Paris, France

S. Levy
Non-Resident Senior Fellow, Brookings Institution, Mexico City, Mexico

H. Bhorat
Professor, University of Cape Town, Cape Town, South Africa

S. Bhalla
Former Executive Director for India, Sri Lanka, Bangladesh and Bhutan, IMF, Delhi, India

V. K. Paul
Member, NITI Aayog, Delhi, India

S. Bery et al. (eds.), *Navigating Challenges for Sustainable Growth*,
https://doi.org/10.1007/978-981-97-7894-2_7

was not able to stop this very favourable trend. Now, I believe that today some uncertainty has built up due to various events. Certainly, the fact that commodity prices have fallen around 2015 and have had a kind of haphazard behaviour since then, the pandemic, certainly the global disorders, and in particular the Ukraine war, but more and more the effects of global warming. And the fact that the policies to mitigate and to adapt to global warming are progressively, or will progressively, be implemented.

This is really creating uncertainty about the process or this new trend in global inequality I was describing, focusing on the big emerging countries like China, India, and others, which may continue to overperform with respect to advanced countries. From that point of view, the process of equalisation may continue, even though it's not clear today whether China, being much above the world mean standard of living, is still contributing to less inequality or is already contributing to more inequality.

But what is really of concern is the fact that low-income countries have not been able lately, practically since over the last eight years or so, to catch up with advanced countries. They are lagging behind advanced countries and of course behind the mean and what is going on in dynamic emerging countries. This is a source of concern because it means we cannot say there is full inclusion. There is a problem with the poor countries in the world, which may not be able to continue the process of development they experienced at the beginning of this century. This is really a problem because it may mean that some exclusion is building up if we are not able to take the right decisions at the national, and most importantly, at the international level, to stop this process and ensure that everybody remains in the group. That poor countries keep catching up with the rest of the world is crucial.

So, this is a source of concern. Another one, of course, is the fact that the various shocks I mentioned earlier will also affect within-country inequality. We know that in many countries of the world, inequality tended to increase substantially over the last decades of the past century. Fortunately, we observed at the beginning of this century a kind of stabilisation, and in several developing countries, even a fall in the degree of inequality. However, it is certainly the case that today inequality is much higher in most countries than it was in the eighties or the nineties.

I'm talking about living standard inequality. Now, in some cases it would be consumption inequality because countries are measuring consumption. And in some other cases, it would be income inequality. But we know that the trends in most countries will be more or less parallel. I don't think it makes a big difference whether we are referring to consumption or income.

This uncertainty building up in the world is also making it less unlikely that there will be another increase in inequality in various countries of the world. These shocks are, most of them, would push toward more inequality. And because the level of inequality is already very high, then all the risks linked to the lack of inclusion, may become even more serious in the future than they are today. And they're already, as we have seen, very serious today.

So, because of that, adjusting to those shocks, developing resilience to these shocks, and making sure that inclusion is a goal, which is very explicitly stated in national and international policies, is of the first importance.

7.2 Speaker 1: Santiago Levy

I would like to share with you a Latin American perspective on social protection in middle-income countries. The basic idea is to point out some lessons that may be useful for other regions of the world, both positive and negative.

Growth is often accompanied by increasing income inequalities and demands for health care. As the population ages, there is a growing need to take care of the elderly and avoid old age poverty. These demands become politically more salient as countries urbanize. Designing effective social protection systems is a real challenge because there is a need to balance budgetary cost, redistribution objectives, and efficiency objectives. This challenge is more complex in countries with large informal employment.

Let me give you an overview of the structure of social protection in Latin America.[1] However, before I do this, a little history is useful. Social protection systems started in Latin America back in the middle of the last century. We imported from Europe the Bismarckian model: workers were going to get social protection through wage employment, and it was going to take the form of a bundled package of benefits. The bundled word is very important because workers were going to get access to health, disability, death insurance, and retirement pensions at the same time. This bundle was going to be financed from a wage-based tax earmarked for these benefits. In addition, workers would be protected from the loss of employment through dismissal regulations and, when employed, through minimum wages, sometimes very high relative to countries' wage distribution.

This model, 80 years later, covers less than half of the labor force of Latin America. After the debt crisis of the 1980s, and as growth resumed in the region, there were pressures to extend social protection to those that were excluded. Countries in the region responded by expanding social protection, but they did not reform the Bismarckian model. What they did is to add on, in a fairly ad hoc manner, a set of programs that are somewhat conflated, but is important to separate. One set of programs were targeted income transfers, like Progresa in Mexico and Bolsa Familia in Brazil, although there are many others in Latin America. A second set of programs were non-contributory insurance programs, which have not received as much attention as they should have but were a central part of the response.

The result is a structure of social protection constructed around workers' status in the labor market. A useful way to think about this structure is to visualize a two-by-two matrix. The columns refer to the provision of insurance against risks: disability, death, illness, longevity, loss of employment, and so on. The rows refer to income redistribution. The columns divide the population in two groups: formal, who have access to the Bismarckian model, and the rest: those informally employed,

[1] For a general discussion of the issues raised in this note, and references to country-specific evidence, see Levy, S. and Cruces, G. (2021). "Time for a New Course: An Essay on Social Protection and Growth in Latin America" United Nations Development Program, Latin American and Caribbean Bureau, Working Paper 24, New York.

unemployed, or out of the labor force, who have access to a set of unbundled packages of services that, on an ad hoc, scheme-by-scheme basis, were being created through non-contributory pensions, health, day-care and similar programs which, very importantly, are paid from general revenues. So, if you are in the right column, your benefits are paid from general taxation. If you are in the left column, benefits are paid by a wage tax that has to be internalized in the contract between the firm and the worker.

The rows of the two-by-two matrix refer to incomes. In the upper row we have the non-poor population and in the lower one those that are poor. Targeted income transfer programs have nothing to do with the columns; they have to do with the rows. Conditional cash transfer programs are for the population in the second row, i.e., for the poor. These programs are not providing insurance; they are providing income transfers, which is very different.

With variations, this structure is present all-over Latin America, although there are exceptions. A notable one refers to health insurance in Brazil. There, health services are completely independent of workers' income or formal-informal status, and this is a very good thing. It is the only country in the region that has done this. In every other country in Latin America, the type of health services that you get depend on whether you are in the right or in the left column of the two-by-two matrix mentioned above. Importantly, note that health services are not being provided through conditional cash transfer programs. These programs provide income; health services, and pensions and disability insurance, are provided through social insurance.

In many discussions about social protection, informality and poverty are conflated, but they are two very different things. It is true that most poor workers are informal. It is not true that most informal workers are poor. If you go back to the two-by-two matrix mentioned before, there are more workers that are informal but not poor, relative to the number of workers that are both informal and poor.

Governments in Latin America spend on conditional cash transfer programs approximately half a percent of GDP. These programs have received a huge amount of attention because there is an abundant literature on their impacts on health, nutrition, and schooling through very careful econometric analysis using difference-in-difference or similar techniques. This has been all to the good because we have learned a huge amount from these studies. But, that said, non-contributory social insurance programs are actually much bigger in terms of both the population that they cover and the budgetary effort that is devoted to them. Some countries can spend between two to three, up to five percent of GDP.

The segmentation of insurance is not a very good scheme because workers transit between the columns, particularly in urban areas, that is, they move between different types of employment throughout their life cycle. This implies that sometimes they receive protection through the Bismarckian scheme and sometimes through non-contributory programs.

What is the result of this overall architecture? I refer here not to any individual program, but to the coexistence of the Bismarckian model with non-contributory insurance programs and conditional cash and other transfer programs. First, protection against risk is erratic because if you change labor status from formal to informal, you may or may not be covered for disability insurance, death, or employment insurance. Second, contributory pensions do not work for the majority of workers. They save for a pension, but many do not get one when they retire because they do not accumulate sufficient numbers of weeks in formality to be able to qualify.

Third, it is true that poverty programs, particularly CCTs, have raised the human capital of the poor. We have a lot of econometric evidence about the impact of Bolsa Familia, Familias en Accion in Colombia, and Progresa in Mexico. This evidence suggests the health status of the poor, their nutrition, and their schooling has increased. However, the increases in their human capital are not translating into better jobs because, despite their increased human capital, they are still being informally employed.

Fourth, the overall scheme changes little the market distribution of income from the after taxes and transfer distribution of income. In most OECD countries, the Gini coefficient of market income is something like 0.47–0.46, but it is brought down to about 0.3 because taxes and transfers play a very redistributive role. In Latin America, the Gini coefficient of market income is slightly higher than the OECD, but not that much higher. The real tragedy is that after taxes and transfers, the change in the Gini coefficient is very small, which basically says that the ability of this whole apparatus, not any individual program, to actually change the market distribution into the post-market distribution is very low. That is the efficacy side of structure.

On the efficiency side, what you are actually doing is, de facto, taxing formality. This is so because the costs of the Bismarckian scheme have to be internalized in the contract between the firm and the worker and, for many reasons, the benefits are less than the costs. My guess is that part of the problem with wage employment in Latin America is that if it is being taxed implicitly through the structure of social protection. In fact, we have a lot of econometric evidence that this in fact happens. And at the same time, informality is being subsidized because governments are providing social protection paid from general revenues, conditional not on being poor, but on being informal.

The same person with the same human capital and abilities is being told: If you have an informal job, your health insurance and maybe your pension will be paid from general revenues. But if you get a formal job, now you and the firm that hires you have to pay for those benefits. Taxing informality and subsidizing informality is not really very good from the point of view of productivity.

What happened in Latin America over the last three decades? There was a large effort to increase public spending in social protection, a reflection of the fact that governments have been deeply concerned about the issue of inequality and poverty. Broadly, for the region as a whole, spending went from seven to about 15 percent of GDP over the course of the last three decades. That is a significant effort. However, despite this effort, we continue to be one of the most unequal regions of the world. Latin America did reduce poverty. But relative to the region's income per capita, poverty is still high. Finally, and although there are many reasons for this, the formal-informal segmentation of workers and firms is one of the reasons (of course, not the only one) behind the stagnation of total factor productivity over the last three decades.

Let me conclude. Latin America pioneered conditional cash transfer programs, but it was not noticed sufficiently that at the same time the region was developing a second-tier system of parallel social insurance. These programs developed in an uncoordinated way, scheme by scheme, as governments responded to social needs by creating a pension program here, a health program there, or a day-care program over there, for various groups excluded from the Bismarckian regime.

This situation has two major problems: One, a dual system of social insurance constructed around worker status in the labor market, a status that fluctuates depending on shocks in demand, technological changes, and idiosyncratic factors. Second, the fact that sometimes insurance and income transfers are conflated. This is an important point to keep in mind when we talk about universal basic income and all that. It is not the same to provide income as to provide insurance, particularly if you want to protect people against catastrophic expenditures.

As it stands today, the architecture of social protection places Latin American governments in a fairly difficult dilemma. From the social inclusion point of view, you want to increase benefits to informal workers so that they have the same benefits as formal ones. From the efficiency point of view, you do not want to keep taxing the formal sector of the economy to subsidize the informal sector because it is not a good idea from the point of view of productivity.

Two more points: Yesterday, there was some discussion about agility and flexibility. This architecture does not respond well to the dynamics of the labor market. It is constructed around the concept of dependent employment, which is the cornerstone of the original Bismarckian model. More than a hundred years after Bismarck, and more than 80 years after this was imported to Latin America, it covers less than half of the population. If we think about more dynamic economies in which there is rapid technical change and people change jobs often over their lifetime, this kind of structure will not work. Middle-income countries from other regions can hopefully learn from both the positive and negative lessons from the Latin American experience and internalize them.

To conclude, as countries develop, they need a vision of social protection, not an accumulation of schemes. There are three key messages from Latin America. First, do not conflate income transfers with social insurance; they have two separate objectives. Second, stay away from constructing the core of social protection around people's status in the labor market. Build it outside the labor market, except for those issues that are directly associated with the behavior of firms, like work risk insurance or unemployment. And third, keep in mind that social inclusion is not achieved through a collection of disconnected social protection programs. Social inclusion requires that all members of society contribute to and benefit from the same social institutions.

7.3 Speaker 2: Haroon Bhorat

If we think about the region and the big picture economic challenges in sub-Saharan Africa, the notion of the relationship between growth and poverty in a region where the majority of the world's poor lives ultimately has to be based on the discussion about Africa's jobs challenge.

And if you're talking about labor demand or a jobs challenge, you have to be thinking about structural transformation. I'll then switch to what I think is a possible roadmap for structural transformation by looking at economic complexity.

Now, these data are well known. But in many ways, when one looks at the trends by three different poverty lines, I've taken the black line with the white dots to represent Sub-Saharan Africa. In this particular case, there is a decline. The first difference does show a decline in poverty levels for Sub-Saharan Africa relative to other regions of the world. But what's very clear is that the rate of reduction is much lower than the rest of the world. That fact means ultimately that Africa's share of the world's poor has risen over time. I think that's a really important background context. These numbers are very well known, but perhaps less well known are the growth-poverty elasticities by region.

So, these are updates from work currently underway within the Africa region, and I'm doing some work with the World Bank team there. It's very clear if you take the median elasticity of poverty reduction to growth. If you think of a one percentage increase in economic growth, what is the reduction in poverty levels? You want that number, that elasticity, to be really high. For sub-Saharan Africa, which is circled on the right-hand side, it's the lowest of all the regions of the world. The growth-poverty nexus is a weak one in sub-Saharan Africa. The ability to convert economic growth to poverty reduction is very weak.

The last bullet is a really important one. The notion that this is not necessarily an Africa issue as much as it is a low-income country or income level issue. So, in other words, if you took a sample of low-income countries or countries with similar GDPs per capita, you'd get similar elasticities. I think that's really important to keep in mind. There isn't an Africa dummy variable sitting there that's significant. Be that as it may, this is a region that at least I and others live in. So it does give you some context for thinking about policy in terms of a region rather than at income levels.

For me, as a micro-econometrician, the notion of jobs, the notion of where employment will arise, is a really critical lifeblood of any economy trying to move through stages of development. Some of the work I've done takes the very simple question looking at changes and projections in the labor force, or the working-age population. Here I've taken the youth, and these are based on the UN population projections. The really important numbers are sitting in the last row. Currently, the youth population of the globe, Africa constitutes 18%, or 19%, of the youth on the planet. By 2100, that number right next to it is going to be 46%. So close to half of the world's youth will sit in this continent. If that doesn't represent the core of a global challenge, a jobs challenge, then I'm not sure what does.

So, in other words, if you think about resilience, if you're thinking about inclusion, thinking about young people or the population of working age, if you look at the numbers to the right, 15% of the population of working age are in Africa. At the moment, that's going to go up to 42%. The majority of the world's youth and the working-age population will actually reside in sub-Saharan Africa. What that means, yes, the macro solution is to say this is a demographic opportunity, but it's a jobs challenge. It's the statistical mirror image. I think that's really important to keep in mind. I do though want to counter against an Africa as homogenous challenge.

So yes, these numbers are true, but if we break it down by country, two-thirds of the youth population right over that growth period that I had to 2100 are accounted for by 10 countries, and you see them there. The usual suspects are there: Nigeria, of course. Some of you may know this, but DRC, Tanzania, Angola, and Niger are the obvious countries just because they're large population economies that are dominant.

Ethiopia is there, but in essence, in many ways, and it does suggest that this is an elephant that you can look at in terms of bite sizes, if I could put it that way. So in many ways, the jobs challenge, in pure scale terms, is a 10-country issue on the continent. And I think that then allows one to look at country policies.

If I dig just a little bit deeper about—yes, we can talk in general terms about the jobs challenge or inclusion in terms of creating jobs, let me just talk about the types of jobs and the employment outcomes that you're seeing in Sub-Saharan Africa.

So what's very clear is that for Africa, employment outcomes are incredibly weak. Africa accounts for 14% of the world's labor force but only 8% of the wage employed. So if you're thinking about inclusion, you better be thinking about wage employment. Yes, second order, I understand, would be informality, but in many ways, wage employment is an indicator for a good job or at least inclusion.

The thing that we often have to orientate ourselves towards is that the majority or disproportionate share of the world's unemployed do not reside in Africa. So we shouldn't confuse those. But the majority of the working poor, because you've got large swaths of workers in agriculture in low-income countries, disproportionately in Africa, that's where you're going to find exclusion in the labor market being concentrated. That is around the working poor rather than unemployment.

South Africa, here's a data point to take away: South Africa is the only African economy where the majority of workers, over 50% of workers, are actually in wage employment. So that really is another way to think about how weak wage employment outcomes are. Effectively, if you exclude the public sector, which sits in wage employment, the effective average private sector wage employment rate in Sub-Saharan Africa is at about 13%. So incredibly weak private sector wage employment outcomes as well.

We know the famous work from Dani Rodrik with Maggie McMillan, the Structural Transformation Bubbles. I've produced these as a mechanism or an entry point for thinking about jobs and growth. What you see in these structural transformation bubbles, I like to call them, is developing Asia on the left, Sub-Saharan Africa on the right. Again, what's happened is a question around whether Sub-Saharan Africa is on a path to adequate structural transformation manufacturing. The manufacturing bubble, if you like, is too small. So effectively, what's happened is, yes, some movement of workers outside of low-productivity agriculture, but the majority of workers in Sub-Saharan Africa are actually going into the informal sector in urban areas and into the public sector.

You can look at it in terms of numbers. You still have the challenge in Asia and SSA. If you compare Asia on the right, I want you to look at the top right figure. You see a massive employment reduction in Asia over this period, 1990 to 2018.

But you still have, uh, Sub-Saharan Africa, with agriculture in Africa, accounting for 35% of all jobs over the period, right? So that's the share of the change. Are there jobs happening in manufacturing? Only 9% of jobs, um, over this period have come from manufacturing in Africa. Compare that to 15% of jobs in Asia. Construction: 35% of jobs have come from construction in Asia over the same period, only 5%. So, weak patterns of structural transformation that are ultimately going to drive inclusion, employment, and certainly wage employment in Africa. We don't really see significant patterns of structural transformation that are going to be employment-inducing.

There is a discussion, of course, parallel to manufacturing about services. And in a separate paper, we've done some work using input–output tables to look at the share of services in manufacturing. And what you do find, the top two lines are Asia. The bottom two lines are Sub-Saharan Africa. And we've taken traditional and modern services; it's very important to separate the two. And it's very clear, even if you think about the role of services in manufacturing, the average combined services used for Sub-Saharan Africa is 18%, compared to 27% for the Asian sample. The OECD average is 22%. The services need to be further embedded into manufacturing as a boost to productivity in manufacturing and employment. That's not happening at the moment.

I strongly still retain that for African policymakers. And that's the next two slides, thinking at a very detailed level about moving from one product that's currently being exported to the one that's closest. To the one that's closest to the current exportable product, and I've produced a few product space maps for Kenya, for Senegal. We've done some work on this. So if Kenya has an apparel and textiles cluster, which they do, which is the green dot (some of you know these graphs, these maps really well), that means that Kenya has a basis upon which to move and guide industrial policy towards building capabilities in apparels and textiles. And the idea, in my view, is this is one very practical way for policymakers to think about building resilience and inclusion-promoting growth at the sub-sectoral level through these product space mappings. I've produced it for South Africa, and we have others for African countries. Ghana's on the left. For example, one very specific example: it took Ghana 30 years to move from being the world's leading exporter of cocoa to finally opening and producing chocolate. That's the idea of moving towards and growing by moving slowly up the value chain, as it were.

There's just a summary of the notions and the ideas that I've brought to bear. I worry about growth, poverty elasticities in the region. I've been looking at them for the last two decades. They have not changed. Part of it is lost opportunities through commodity supercycles, which we can talk about. But ultimately, as we look forward, the jobs challenge is massive in this region. And if you think of the spillover effects for the North, in terms of not resolving a jobs challenge, those are obviously the migration crisis being a clear one. Finally, when we think about a different way to do economic policy that sustains growth in Africa, I suggest that product space and analytics may be one route in.

7.4 Speaker 3: Surjit Bhalla

I'm going to talk about three topics and all related to the headline: 'Data is the New Oil.' It was also the old oil, and it is political. Whether it should be or not, I don't think it should be, but that's what I hope to discuss.

Measurement and evaluation of inclusion is a central part of any policy on inclusive growth. In India, questions on what has happened to poverty, inequality, and employment are extremely controversial. I'll go through the data from 1983 to 2022 on government data, non-government data, etc., to try and substantiate that there is a real problem with interpretation of data and perhaps with the data itself.

A simple definition item to keep in mind is that inclusion means at least an equal sharing of the proceeds from growth. Inclusive growth is when, regardless of the indicator, e.g., share of bottom 40% in consumption or income (let us call it X), this share should grow at least at the same rate as that pertaining to average. The average can grow fast, or slow, or negative—inclusion means that X should be higher than the average.

Has World Growth Been Inclusive?

Part of the controversial nature of inequality is that there is a belief that world inequality has deteriorated considerably. However, the data shows that at a world level, there has been considerable inclusive growth. At an individual, and world level, this is easily verified by the fact that two very large, and very poor countries, India and China, have had per capita income growth far exceeding the world average, and far exceeding the average for the Advanced economies, and this has happened over the last 30 years—and continuing.

What the China-India growth data underline is that inequality *has* to have improved and improved dramatically. India and China's average growth rate for the last 30 years has been 5.9% per year. These two large economies account for something like 40% of the global population, and they were the poorest in the world in 1980.

The two economies have grown; their average income has increased at a 5.9% rate, and the world average is something like less than 2.3% because the world average includes the 5.9%. So, inequality has to have improved. When I first presented these results back in 2002 in a book called '*Imagine There's No Country*,' it was met with considerable skepticism, not to mention opposition.

High Growth in India and China

Next, an important background to what has happened to world inequality is that the big change over the last 40 years, perhaps the biggest change along with a decline in fertility, is the rise in average levels of education, especially amongst the poorest countries of the world.

This will have, and has had, expansive effects. In a study in 2017 (The New Wealth of Nations), I showed that the Western world in 1992, had the same number of college graduates, that is the flow, not the stock. The same number of college graduates, about 80 million, graduating in 1992 and in 2016. And these numbers have really jumped since then. The Western world went up to 124 million, and the rest of the world went up to 263 million.

The fact that the supply of richer salaried workers has moved outward has had predictable effects. One of the consequences is that based on NSS data on wages since 1983, we observe that unskilled workers' wages have expanded at an annual rate of 5.2%, and the skilled, which includes the salaried workers (in India, we have a classification: are you salaried, are you casual hourly wage worker, or are you self-employed?) The casual and salaried workers today account for approximately 25% each in India. And as shown in Table 7.2, and especially since 2011, the casual workers' wages have really gone up, whereas salaried worker wages have stayed nearly the same.

Table 7.1 Supply and wages of college educated workers, 1960–2016

	1960	1973	1980	1992	2000	2016
Real wages[1]						
Some college	-	938	880	925	1028	1117
College	-	1125	1041	1344	1512	1640
Completed college education *(in millions)*						
US	10.1	19.2	27.4	40.3	45.3	59.9
West	16.5	34.4	50.0	79.1	96.9	124.2
Rest	7.1	19.1	32.5	80.7	136.2	263.1
Gap in supply of college educated workers[2]						
Ratio - (West / Rest)	2.3	1.8	1.5	1.0	0.7	0.5
Percentage difference[3]	132.4	80.1	53.8	-2.0	-28.9	-52.8
Median inflation in the West (in %)	1.6	8.8	12.3	3.2	2.7	0.4

Source Economic Policy Institute; [available at] http://www.epi.org/data/#?subject=wage-education

Notes 1. Data sourced from Bureau of Labor Statistics, in 2016 prices

2. Aggregate of individuals with completed college degrees (from Barro-Lee Data) is totaled for the West (advanced economies) and the Rest (all other economies)

3. Percentage difference calculated as 100*(West - Rest)/Rest

Table 7.1 from Surjit S Bhalla, The New Wealth of Nations

Table 7.2 Real wages for skilled and unskilled workers

Year	Casual (unskilled) worker	Skilled (salaried worker)	Worker (either skilled or unskilled)
1983	41	132	73
1993	56	195	102
1999	81	268	154
2004	90	263	154
2011	137	341	220
2017	174	319	247
2018	183	313	251
2019	179	305	246
2020	186	305	247
2021	213	313	265

Source NSS & PLFS data, 1983–2021; authors computations
Note In 2011–12 prices, Rs per week

Table 7.3 Mean years of education

	Ages 15–64		Youth ages 15–24	
Year	Women	Men	Women	Men
1983	0.6	1.2	0.9	1.5
1993	2.7	4.7	3.8	5.3
1999	3.2	5.2	4.5	5.8
2004	4.2	6.5	6.1	7.4
2011	5.5	7.5	7.9	8.7
2017	6.4	8.2	9.2	9.6
2018	6.7	8.4	9.4	9.8
2019	6.8	8.5	9.6	9.8
2020	6.8	8.5	9.6	9.8
2021	7	8.7	9.8	9.9

Source NSS & PLFS data

Also Inclusive—Expansion of Education in India

Now, one of the biggest stories in the world, and in India, is what has happened to education. I've given the numbers for 15 to 24, which is the youth. There is complete equality between men and women in education in India today. The 15 to 64 age group has a legacy effect or generational effect, so it's much more meaningful to look at 15 to 24 or 15 to 29. And this is really quite striking. There are many other pieces of evidence about gender equality in education in India.

There are more women in college in India today than men. Oxford University reached, after a thousand-year history just a few years back, the milestone of there being more women in college than men. This is a real, genuine revolution, which will have consequent effects on wages and occupations in the future.

One other thing to note about gender equality in India. Female pilots in India are the highest in the world at about 15%, versus a 3% world average. STEM enrollment in India is something close to about 42%, whereas in the US, it's something close to 31%.

Income Distribution in India—No Official Data

Official income distribution data for India are not available. Consumption inequality data are available, and the various surveys that have come out since 2011 show that consumption inequality has declined post-2011, though we await the results of the 2022–2023 Indian consumption expenditure survey.

Wage Discrimination and Inequality in India

Discrimination can take many forms. Inclusion, non-inclusion can take many forms. One less-studied aspect of wage inequality is that between communities, e.g., Hindu–Muslim. And what you have is that the median real wage of Muslims relative to Hindus is higher, but now it's about equal. So, no evidence that there is wage discrimination, and it includes effects of the endowments, etc., via the Oaxaca-Blinder decomposition (Table 7.4).

Further, in the US, Black-White earnings differentials have been much studied. Since about the early 1970s there has been a slight deterioration. So this has to be kept in perspective that we have an example from the most open democratic country in the world as to what has happened there, and contrast it with the nature of inclusive growth in India.

Table 7.4 Inclusion via no discrimination

Year	Male median real wage		Male median real wage ratios (%)		
	Muslims	Hindus	Muslim/Hindu	Ratio by oaxaca-blinder decomposition	US Black/White
1983	48.4	40.8	119	110	81.7
1993	72.3	63.7	114	105	80.2
1999	105.1	93.6	112	103	80.4
2004	101.6	97.7	104	100	80.1
2011	145.9	150	97	93	77.8
2017	192.1	192	100	95	74.7
2021	238	228	104	101	78.6

Source NSS & PLFS data; EPI data for USA
Notes In 2011–12 prices, Rs per week

Trends in Absolute Poverty

A recent MF working paper concluded that India has been successful in eliminating extreme poverty (less than 1% of the population in 2021–22; Bhalla-Bhasin-Virmani April 2022.) This is not the conclusion of the gold standard of poverty estimates, the World Bank. The important question arises: how did we arrive at the conclusion that extreme poverty has been eliminated to less than 1%, whereas the World Bank thinks it's something like 10–15%?

The working paper documents that World Bank *assumptions* bias their results. The World Bank uses a very old-fashioned 30-day recall period method to measure consumption, and therefore poverty (defined as per capita consumption less than PPP$ 1.9 per day per person). Starting in 1999–2000, the survey authorities moved toward measuring consumption according to a Modified Mixed Reference Period (MMRP) basis. The big difference between the World Bank method and the official Government of India method is that the latter measures consumption in a more elaborate manner e.g. food and perishables are measured on a weekly recall basis, consumer durables on a 365 basis. Just this modification led to the poverty estimate in India to be 13% in India, compared to 23% obtained from the World Bank uniform recall method. And extending the MMRP to 2021–22, and incorporation of food subsidies, led Bhalla-Bhasin-Virmani to conclude that poverty in India was less than 1 percent of the population.

Counter Opinion on Inclusion by International Scholars

In conclusion. I want to point out some counter-opinion on inclusion in India. Despite a 50% growth in real average consumption post 2011–12, international and domestic scholars (IDS) maintain that extreme poverty in India has stayed constant at around 20%. That is, zero consumption increase for the bottom fifth of the population. Second, UN employment projections show that the age group 15 to 64 will expand by a hundred million over the next decade, approximately 10 million a year. Yet, IDS maintains that more than a 100 million *jobs* are needed over the next decade, with one prominent scholar estimating the need to be 200 million.

Official PLFS data shows that the female labor force participation rate is between 28 to 36%, depending on the employment definition. IDS scholars maintain, on the basis of CMIE data which they use instead of PLFS, that the Indian female labor force participation rate is around 8%, the lowest in the world and well below Yemen. And this passes as 'scholarly' work.

And that's the question with which I want to conclude: do such debates happen elsewhere? At this table, and at this conference, we have representation from around the world, major scholars who have worked on poverty, inclusion, and income inequality. All I am asking is whether such non-facts debate happens elsewhere? In my reading, it doesn't.

7.5 Expert Comment: Vinod Kumar Paul

I would like to touch upon two themes related to the major part of the discussion in this session.

First, India has seen a major reduction in multi- dimensional poverty. Compared to 2015–2016 vis-a-vis 2019–2021, in this five-year span, multi-dimensional poverty declined from 25% to 15%. And, the most dramatic decline was not in urban areas, but in rural areas, from 33% to 19%, almost a one-third decline. As many as 135 million individuals, citizens of India, came out of multi- dimensional poverty during this relatively short period of time. We can expect achieving SDG target 1.2 much ahead of the 2030 timeline.

The World Inequality Report shows that globally, the top 10% of people have 50% or more of the income. And, the bottom 50% typically have less than 10% of the income. My question is, what is the norm for an inclusive, just equitable society going forward? Having learned from the past, going into the 21st century and beyond, what should be that mix of income distribution across population quintiles?

My second point is in relation to the role of financial shocks related to healthcare that trigger impoverishment, inequalities and poverty. The impoverishment due to health spending in India in 2015–2016 has been estimated in a study to be 5.1%. It is estimated that over 32 million individuals are pushed into poverty every year due to catastrophic health expenditure. Our out- of-pocket expenditure between 2015–2016 and 2019–2020 has declined from 65% to 47%. The global average of out-of-pocket expenditure is 17%. Can we do better globally with the help of G20 and international community? We endeavor reduce out-of-pocket expenditure for health further.

You may be aware of the Ayushman Bharat PMJAY flagship scheme of India. This is the world's largest publicly-funded health assurance programme which provides free care for hospitalisation to over 600 million citizens of India. In the course of about five years, we have seen 54 million hospitalisations; typically, about 50,000 per day. And we believe on the basis of the estimates available to us that this scheme has saved $12 billion US dollars in out-of-pocket expenditure to the citizens of India since late 2018.

The major reason for out-of-pocket expenditure is outpatient expenditure on medicines and diagnostics. We're trying to make available over a hundred drugs through our primary healthcare system, which has seen a transformation with the institutionalisation of more than 150,000 Health and Wellness Centers in a span of less than four years, from 2018 to 2022.

Being the pharmacy of the world, India is in a position to expand contribution of affordable medicines for the world. We also believe that we can offer inexpensive, affordable diagnostics. In the course of Covid, almost 300 new diagnostics came from India in a matter of a few months.

We look forward to your ideas on how to reduce inequalities, and how to achieve the Universal Health Care SDG target.

Chapter 8
Keynote Address by Nandan Nilekani

Nandan Nilekani

8.1 Nandan Nilekani

Thank you very much, Suman. It's really great to be in this august gathering. I'll speak today about digital public infrastructure, but in the context of a fragile world and how you need to be agile in a fragile world. I think we all know the situation. The world is getting hotter, 1.5 degrees, two degrees. The world is getting older, and you can see that in the next hundred years, there's going to be a lot more old people. And we know that relationships are getting colder now. All friend shoring, you want to start geopolitics. Different countries impact that on the supply chain. So, it's a very complicated situation that is emerging.

Now, obviously, I think you have figured out in this gathering that India is uniquely placed. It's a young democratic nation with a lot of potential. And many estimates on how it's gonna be number two, number three, whatever in the world. But actually, if you look inside India, we have historically had huge variations in cultures, markets, industrialisation, and regulations. With a country with so many languages, you can't even talk to another Indian. It's often difficult because they don't have any common language. We had very many micro markets. States had different tax laws. People would have a warehouse in every state because they had to have a way to service each state. Industrialisation across the country is very different, and of course, many regulations at the state level. So there was not really a single market until recently. And I'll talk about how technology has enabled this creation of a single market with all its consequences on the economy.

Now, what I think is happening, and it's maybe a 10–20 year journey, but fundamentally, India is moving from an offline, informal, low productivity set of micro economies to a single online, formal, high productivity mega economy. That's a 20-year process, but I think you can see the beginnings of that change. And that's one

N. Nilekani (✉)
Chairman and Co-founder, Infosys Ltd., Bangalore, India

Founding Chairman UIDAI (Aadhaar), Delhi, India

S. Bery et al. (eds.), *Navigating Challenges for Sustainable Growth*,
https://doi.org/10.1007/978-981-97-7894-2_8

of the reasons which is the root of the transformation happening in the country. A lot of this is enabled by what we call digital public infrastructure, which is public digital infrastructure at population scale meant for everyone and either built with public funding or enabled by the public through regulation and policy. Now, these are really building blocks, and these building blocks have been designed over the last 15 years. Each building block does one thing, but then each sits with others, and the interoperability of those building blocks creates a lot of real innovation.

Now, digital public infrastructure is not a new idea. If you look at the original internet, it was funded by the US Department of Defense. The worldwide web was designed at CERN, which was funded by European countries. The Mosaic browser was built by a grant from the National Science Foundation, so fundamentally, the digital public infrastructure, the original one, is the internet built by public money. But then it had a set of protocols on top of it, which led to market innovation, and that led to the rise of Google, Facebook, and so on. Similarly, GPS was again funded by defense in the US. It answered the question, "Where am I?" That became the basis for maps that became the basis for ride hailing. So we have seen this movie before where you can invest in digital public infrastructure and build on top of that through innovation. What has happened in India has taken that idea forward in many new areas and created a whole new infrastructure, not only for basic technology but to enable equitable growth, which cuts across regions and so on.

Now, an example of that is how DPI was used in financial inclusion in India. Now, as the BIS recently reported, India was able to do in nine years in financial inclusion what would normally have taken 47 years. In other words, the acceleration of financial inclusion where the number of people with bank accounts went up from 20% to 80% in nine years was essentially lubricated or done with technology. And this is something which has had a huge impact on the country in terms of creating a more inclusive model. Similarly, what has happened with cash? In 2016, Indians were mostly using cash payments. In six years, we now have the largest volume of digital payments in the world. About 40% of digital payments in the world are done in India today. UPI, which is a very low-cost, high-volume transaction processing system run by the National Payment Corporation from India, does 9 billion transactions a month. It has about 350 million users and about 50 million merchants where you can make digital payments. This has come from nowhere and suddenly become the world's largest digital payment infrastructure. Fundamentally because the philosophy and the way of thinking about it was, "How do we create a national digital public infrastructure?"

And Subbu was kind enough to talk about Aadhaar, which I led the project for five years. Essentially, we had to solve the problem of how do you give people without any ID an ID. Give people without a birth certificate a starting ID. And today, 1.3 billion people are on the ID platform. And this is something which has become foundational for everything else. And this was the root of creating a single source of truth, reducing duplication, and so on. And this ID is used for online identity verification up to 80 million times a day for the transaction volume just on ID. 80 million times a day that it's used by somebody somewhere to verify that he's who he claims to be. So, think about the volumes you're talking about. You're talking about payment transactions of 9 billion UPI transactions a month, 80 million Aadhaar authentications a day, and so on and so forth. So these are really very large scale, 24 by 7, real-time, quick response systems that have been built in India. And these, I talked about the fact that DPIs do one thing at a time. So each of these things allows other things to happen because they're all built using interfaces, protocols, programming interfaces, and so on.

So, Aadhaar was the basis for identity, but on top of Aadhaar was built a capability called KYC, or "Know Your Customer." How do you, how do you, because for many things like opening a bank account or getting a mobile connection, you need to know the customer, otherwise you can't do it. And we built e-KYC as a digital way of doing this KYC, which you can do in two minutes.

Now, that led to two revolutions. One was the banking revolution I referred to earlier, which is thanks to the Prime Minister's Jan Dhan program, which was launched in 2014. 700 million new bank accounts got opened with Aadhaar KYC. So the KYC enabled people both to get an ID and verify that ID in real time to open a bank account. And that's how, and that is used not only for cash transfers, it's also used for non-cash transfers. So in a PDS system, which is a national distribution of basic amenities like rice and so on, all authentication is done with Aadhaar. So this was on the benefit side, but the same thing happened on the mobile side.

And in 2016, when Jio was launched, Jio essentially was a new mobile network, India's first truly 4G network. And that network had the goal of getting to a hundred million customers in six months, which means they had to enroll or onboard 1 million customers a day. And the only way they could do that was to use Aadhaar e-KYC for mobile connections. That essentially transformed the mobile industry. The mobile industry went from one gigabyte (GB) of data per month consumption to one GB a day, a 30x increase in data consumption after the launch of Jio. And similarly, the smartphone penetration went up to about 70%.

So essentially, what these three things did was lay the digital foundation for equitable participation of everybody. Everybody had a digital ID, digital ID could be used anywhere in the country for online authentication, so it made it friction-free for you to travel in the country. Digital ID gave you KYC, KYC gave you a bank account. Digital ID gave you KYC for mobile, gave you a smartphone connection. So everyone could get a smartphone connection, a bank account, and an ID, and they're good to go going forward. Essentially, that's what DPI does. It allows you to mix and match things and create all kinds of solutions on top of that.

Similarly, another great initiative of digital, which is by DigiLocker, is a single repository for all your digital documents. So this is a single market for credentials. And today, it has about 180 million people using it, and they keep all their documents there. There are about 5 billion documents stored here. So a person would keep his driver's license, his Aadhaar details, his vaccination certificate, his vehicle registration, and they don't need to carry it anywhere. They can just show it on the phone, and it's digitally certified that it has come from the source, so it eliminates fraud. This again is like a national market for credentials. You can get your credentials in Bihar, migrate to Delhi, and use your credentials there. All this helps in reducing transaction costs, brings in efficiency, and so on.

And then UPI, I talked about UPI, which was the payment system, which does 9 billion transactions a month, but it's also changed the way India does merchant payments. In India, POS machines took about 75 years to reach 6 million point of sale machines. Now, in a matter of four years, we have 50 million merchants where you can pay using a QR code. And these QR codes are just stuck there in front. These are coconut vendors selling coconuts who have a couple of QR codes, and I just have to point my phone at the QR code and make a payment. And this has essentially dramatically transformed business transactions or retail transactions. And you also have things called the sound box where you can hear that the payment has been received, which improves the productivity of a small merchant because he doesn't have to handle cash. You just hear that the money has come. So all these innovations have happened to essentially reduce transactions in business transactions. And this again is happening at a huge scale.

And then, of course, the GSTN, which was another great initiative, has essentially created a backbone for a single market. So think about it this way: that identity, bank accounts, and mobile phones operating anywhere in the country created a single market for services and people. And GSTN created a single market for products. And what's important is that it has not only led to an increase in revenues because of improved compliance through better revenue collection, but it also provides the data for what we'll come to later, which we call digital capital. The way the GSTN is designed is actually a company which is jointly owned by the Indian government and the states, which operates our tax system. So India went to a simplified tax system for everybody for indirect tax, and also before this, different states were at different levels of tax systems. Now everybody is at the same level, everybody uses the same system, everybody files their returns online. The same thing happens in income tax. We have gone to a completely online income tax system. So the combination of both indirect tax and direct tax being completely digital has had huge benefits, including raising the tax revenues of the country.

And then, of course, the logistical improvements on the highways. GST has created single markets. You don't have to have a warehouse everywhere. You can have one warehouse. FASTag is the RFID tag, which every truck and car today has. So when they pass through a toll gate, they automatically debit the wallet attached to the FASTag.

This system alone does 2.5 billion transactions a year, and it has had a dramatic impact on efficiency and productivity on the roads because you don't have to stop. And it's expected that this year it'll be about 8 billion transactions.

Now, all these things have also established a way to strengthen a company's public finances, which I think you'll all agree is very important because tax revenues have gone up. Tax revenues in India for the last 10 years have risen faster than GDP growth, and that is essentially because of technology, compliance, and formalisation. More and more people are joining the formal economy. Similarly, that's on the revenue side. On the expenditure side, all welfare spending is done through Aadhaar- linked bank accounts, which makes sure that the money goes to the right person. Cumulatively, since inception, India has transferred $210 billion directly into people's bank accounts. That has made a huge impact on delivering benefits without any leakages. Moreover, the whole PFMS, the financial management system, is linked with Aadhaar, so when money is sent, you can drill down to granularity as to whom it was sent to. That dramatically reduces leakages and corruption in the system.

And then, of course, in the case of credit, I'll come back to that. So basically, these things actually have a material impact on the economy. Then, the digital public infrastructure also makes markets more equitable and competitive. The latest initiative in India is called ONDC, Open Network for Digital Commerce, which unbundled commerce. E-commerce historically has been a single company where you place the order, which has a list of suppliers, and which delivers to you. ONDC does unbundled commerce. So I can order from any app, from any supplier, and have it delivered by any logistics company. This is really opening up commerce so a small guy in a small town can list his products using the ONDC protocol, and anybody can buy from that person. We think this will democratise discovery, reduce transaction costs, and also create a much healthier, competitive e-commerce market so that it's not like a winner- take-all with just one or two players.

A good example of that is a union of auto-rickshaw drivers with the three-wheelers in Bangalore. They are using a protocol, and they are just getting the rides discovered on the platform, but they don't pay anybody in between. The transaction is between the consumer and the driver. So he gets 100% of the money, not 70% of the money. There's nobody taking a cut. There's no aggregator taking a cut in between. This is already doing 80,000 transactions a day, which means 80,000 times in a day, some person in Bangalore is using this platform to get an auto-rickshaw. And you can imagine this can then be done for taxis. Fundamentally, it's restructuring the way mobility is done in India.

And then, of course, we have AI initiatives like Bhashini. Bhashini is an initiative to create a complete AI base for all 22 major Indian languages. And this is all being built as an open stack so that anyone can use it. It's for speech-to-speech, text-to-speech, text-to-text, etc. And it essentially means that every Indian will get access to knowledge in their language of choice, verbally spoken to them, through speech. This dramatically improves access because I was talking to you about the fact that we have so much diversity. This initiative will essentially allow everyone to participate. So, in time, I'll make a payment on my phone using UPI in Bhojpuri, and that's coming in the next year. So these are all fundamentally changing access to technology for millions of people.

What we are saying is DPIs also create data as a byproduct because, as we know, every digital transaction has data as a byproduct. But in the Western world, that data is captured by large platforms who then use it to sell ads to you. The data is not with you. In a small autocratic world, data is used as a surveillance mechanism. The Indian model gives data back to you, and we call that digital capital. Just as we have historically thought of land, labor, and capital as assets, digital capital is the latest form of capital. But you can only unlock digital capital if you can make it accessible to people. That's what Indian architecture has done. We have an architecture in India called the Account Aggregator, sponsored by the Reserve Bank of India, which allows for all financial transactions. An individual can get access to his or her own data and then use it to get a loan or buy a mutual fund. So the fact that we are unlocking digital capital is another huge driver for inclusive economic growth. A young person gets an ID, uses the ID to open a bank account, uses UPI to get payments, then based on their transaction history, they get access to loans, build a credit history, get business loans, then use an online education platform to learn skills, and so on. Basically, you can create a pathway for individuals and businesses to get access and join the formal economy.

We are seeing the same thing happening in credit today with digital capital so that people can give a history of their financial transactions. e-KYC, UPI, and small loans are exploding in India. This is the democratisation of credit. Historically, credit went to large entities because they were the only ones who had data to prove that they were worthy of a loan. But now millions of small people are going to get credit. Credit is going to go both to the buyer and the seller. Consumers will get credit based on their history. Suppliers will get credit based on their history. You're essentially going to turbocharge the economy by giving credit to both buyers and sellers. And that's part of the reason why you're going to see serious economic growth here.

And this digital capital can also be accessed by countries, for example. The Fastag I talked about is essentially used for efficiency at tollgates, but the byproduct of that is there's no leakage of money at the tollgates. All the money is collected because it's all digital, and the tollgates become much more bankable. Now, we are seeing that the government, NHAI, is able to sell the toll gates because the buyers are sure of the revenues and can reinvest that money into new roads. The recycling of capital for building infrastructure will also happen because the digital capital will fund new physical infrastructure. You can sell or securitise the old assets.

Similarly, all this will enable many AI applications. I saw you had a discussion on AI earlier, but I won't get into details. Fundamentally, all this will also enable the application of AI for various solutions. We can talk about that later. So, what this does is it provides the basis for formalisation. Formalisation is a challenge in every society. How do you create a formal economy? What's in it for the person? If a person is outside the system, why should they join? They might just get stuck with bureaucracy and taxes. Now, we are offering a new bargain: if you join the system and we simplify that with technology, you can then use your digital capital to get ahead. If I'm a small business relying on informal credit and I enter the system, I'll get formal credit. We're creating the incentive for people to join the formal system. That's what I meant by formalisation. This is going to happen over the next 20 years.

Digital infrastructure is only as useful as the people who build solutions. As I said, the internet was there, but you think of the internet as using Uber, Facebook, or Google. Similarly, we need to create a way for these building blocks to be reconfigured to create solutions. India is very well placed because it has a huge entrepreneurial class which is leveraging this technology to build innovative solutions.

NASSCOM, the Indian IT Services, laid the foundation for this. It's a $227 million industry employing 5 million people, and that has provided the technology and the talent for making the DPI possible. Of course, there's a huge startup system on top. Just to give you a sense of the scale, there were 1,000 startups in India in 2016. Today, there are 115,000 startups. Of the 115,000, about 30,000 have gone bankrupt. But it doesn't matter. Innovation is happening at a crazy scale. The innovation ecosystem is built on top of the digital public infrastructure we have.

And finally, we talked about agility. In this new world, agility is the ability to respond in policy at scale. Digital public infrastructure is the key to that. How do you respond quickly to changes in the world? And also, how do you balance between regulation and innovation? We do that by embedding policy as code. So, if you look at ONDC, for example, the policy of the ecosystem is in the code. Or if you look at UPI, where there are multiple banks and multiple apps, the policy for how they operate is in the code of those platforms. This allows us to be much more flexible and reduce the time between thinking of policy and implementing it.

In times of crisis, DPIs actually elevate your response.

For example, the entire vaccination programme of India was done on a common platform built by the government called CoWIN. Everybody could get vaccinated anywhere. They received a digital vaccination certificate they could keep in their digital locker. That took off. And then, using the plumbing, you could transfer money. So, $4.5 billion was transferred during the COVID pandemic into bank accounts of 160 million beneficiaries. You could make that decision in real time. If you want to give money to these people, you can do it. Of course, everybody uses this platform now, so it's interesting to see how it's evolving.

Now, we talked about climate and green growth. Digital public infrastructure can also be used for both adaptation and mitigation. For example, if you have the plumbing, you can make anticipatory climate financing for improving resilience. You can use it for giving emergency money for a natural disaster. You can use it to prevent forests from being cut. All that requires plumbing. Having agnostic plumbing allows you to also use it for climate. ONDC allows you to have a circular economy. ONDC can now have a reverse logistics specialist who takes the stuff back so that it's recycled, and so on. And similarly, as you go from monolithic power generation to thousands of batteries in thousands of cars bumping up power using a feed-in tariff, you need infrastructure that's interoperable for that. You can build all kinds of things for that. I won't go into details, but fundamentally, even for the climate transition, you'll need to think through what kind of digital infrastructure is required at a population scale.

Moreover, finally, let me say, a DPI approach requires deep conviction, not deep pockets. This isn't about money. We spend all our time arguing about trillion dollars for this and that. This doesn't need any of that. This just costs a few billion dollars, but it has a massive payoff. The ROI is very high. India spent a billion and a half on Aadhaar and has saved $27 billion. You can't get a better ROI. Even VCs would like to see that kind of ROI. When you build this stuff, it's built for high volume, low cost, sachet- sized transactions. It's built to innovate on top. And as I said, the policy can be embedded in the code. Digital capital is key for this. I don't have time to go into details, but fundamentally, I believe green growth in an uncertain world needs a new agile approach, and every nation needs such a digital transformation.

Thank you very much.

Appendix 1
Conference Outline

Friday, July 28	
9.15 AM–9.30 AM	Welcome address: CEO, NITI Aayog and President, IDRC
9.30 AM–9.45 AM	Goals and Process—G20 India Sherpa, VC NITI Aayog
9:45 AM–11:15 AM (90 min)	**Session 1—Energy, Climate, Growth** **Chair:** Jayant Sinha, Member of Parliament & Chair of Parliamentary Standing Committee on Finance (India) *Discussion leaders* • Robert Stavins, AJ Meyer Professor, Energy and Economic Development, Harvard University, Cambridge MA (USA) • Jessica Seddon, Senior Fellow, Yale Jackson School of Global Affairs (USA) • Arunabha Ghosh, CEO, Council on Energy, Environment and Water (CEEW), New Delhi (India)
11:15 AM–11:35 AM	**Tea/Coffee Break**
11:35 AM–1:05 PM (90 min)	**Session 2—Technology, Policy, Jobs** **Chair:** Sachin Chaturvedi, Director General, Research and Information System for Developing Countries (RIS), New Delhi (India) *Discussion Leaders* • Paul Samson, President, Center for International Governance Innovation (CIGI) (Canada) • Albert van Jaarsveld, Director General, International Institute for Applied Systems Analysis (IASA) Laxenburg, (Austria) • Debjani Ghosh, President, National Association of Software & Services Companies (NASSCOM) New Delhi (India) *Expert Comment:* V. K. Saraswat, Member, NITI Aayog (India)
1:30 PM–2:00 PM	Introduction by BVR Subrahmanyam, CEO, NITI Aayog (India) *Keynote address:* Nandan Nilekani, Chairman and Co-founder, Infosys and Founding Chairman UIDAI (Aadhaar)
2.00 PM–3:00 PM	**Lunch**

(continued)

S. Bery et al. (eds.), *Navigating Challenges for Sustainable Growth*,
https://doi.org/10.1007/978-981-97-7894-2

(continued)

Friday, July 28	
9.15 AM–9.30 AM	Welcome address: CEO, NITI Aayog and President, IDRC
3:00 PM–4:30 PM (90 min)	**Session 3—Growth implications of a fractured trading system Chair:** Peter Drysdale, Emeritus Professor of Economics and Head of the East Asian Bureau of Economic Research, Australian National University, Canberra (Australia) *Discussion leaders* • Alicia Garcia-Herrero, Chief Economist, Asia Pacific Natixis (Spain) • Nagesh Kumar, Director and Chief Executive, Institute for Studies in Industrial Development (ISID) (India) • Otaviano Canuto, Senior Fellow, the Policy Center for the New South (Brazil) *Expert Comment:* BVR Subrahmanyam, CEO, NITI Aayog (India)
4:30 PM–04:45 PM	**Tea/Coffee Break**
4:45 PM–06:15 PM (90 min)	**Session 4—Reshaping Global Finance for Sustainable Growth** **Chair:** N. K. Singh, Chairman, Finance Commission and President, Institute of Economic Growth (India) *Discussion Leaders* • Hanan Morsy, Deputy Executive Secretary and Chief Economist, United Nations Economics Commission for Africa, (Ethiopia) • Tao Zhang, Chief Representative for Asia and the Pacific, Bank of International Settlements (BIS) (China) • Poonam Gupta, Director General, National Council of Applied Economic Research (NCAER) (India) *Expert Comments:* Manjeev Singh Puri, Former Ambassador of India to the EU, Distinguished Fellow, The Energy and Resources Institute (India)
6:15 PM–7.30 PM	**Virtual Discussion and wrap up of Day 1** **Chair**: Ashima Goyal, Professor, Indira Gandhi Institute of Development Research (IGIDR) *Discussion Leaders* • Robert Lawrence, Albert L. Williams Professor of International Trade and Investment, Harvard University Cambridge MA-*Technology, Policy, Jobs* [Virtual](USA) • Homi Kharas, Senior Fellow, Center for Sustainable Development, Brookings Institution Washington DC– *Multilateralism: Geopolitics, Governance and the global commons* [Virtual] (USA) • Kapil Kapoor, Regional Director-Asia, IDRC, New Delhi – *Summary of the first day*
Saturday, July 29	
9:30 AM–9:45 AM	Opening Remarks: VC, NITI Aayog

(continued)

(continued)

Friday, July 28	
9.15 AM–9.30 AM	Welcome address: CEO, NITI Aayog and President, IDRC
9:45 AM–11:15 AM (90 min)	**Session 5—Multilateralism: Geopolitics, governance and the global commons** **Chair:** V. Anantha Nageswaran, Chief Economic Adviser, Ministry of Finance, Government of India, New Delhi (India) *Discussion Leaders* • Jean-Louis Arcand, President, Global Development Network (GDN), Geneva (Switzerland) • Mari Pangestu, Former Managing Director, Development Policy and Partnership, World Bank (Indonesia) • Ram Madhav, President, India Foundation (India) *Expert Comment:* Ramesh Chand, Member, NITI Aayog (India)
11:15 AM–11:35 AM	**Tea/Coffee Break**
11:35 AM–1:05 PM	**Session 6—Adjustment, Resilience and Inclusion in an Uncertain World** **Chair:** François Bourguignon, Chair, GDN Board; Professor Emeritus, Paris School of Economics, former Chief Economist, World Bank (France) *Discussion leaders* • Santiago Levy, Non-Resident Senior Fellow, Brookings Institution (Mexico) • Haroon Bhorat, Professor, University of Cape Town, [Virtual] (South Africa) • Surjit Bhalla, former Executive Director for India, Sri Lanka, Bangladesh and Bhutan, IMF (India) *Expert Comment:* Vinod Paul, Member, NITI Aayog (India)
1:05 PM–2:15 PM	**Lunch**
2.15 PM–3.45 PM	**Round-table: How must the G20 evolve? Chaired by** Jean-Louis Arcand, President, GDN and Amitabh Kant, G20 Sherpa of India
3:55 PM–4:00 PM	Closing Remarks by BVR Subrahmanyam, CEO, NITI Aayog (India)

Appendix 2
About the Organisers

NITI Aayog

The NITI Aayog serves as the apex public policy think tank of the Government of India, and the nodal agency tasked with catalysing economic development, and fostering cooperative federalism through the involvement of State Governments of India in the economic policy-making process using a bottom-up approach. NITI Aayog is developing itself as a state-of-the-art resource centre with the necessary knowledge and skills that will enable it to act with speed, promote research and innovation, provide strategic policy advice for the government, and deal with contingent issues.

Learn more at www.niti.gov.in.

International Development Research Centre (IDRC)

As part of Canada's foreign affairs and development efforts, the International Development Research Centre (IDRC) invests in high-quality research in developing countries, shares knowledge with researchers and policymakers for greater uptake and use, and mobilizes global alliances to build a more sustainable and inclusive world. IDRC's investments aim to achieve the United Nations' Sustainable Development Goals (SDGs). They focus on five key areas including climate-resilient food systems; global health; education and science; democratic and inclusive governance; and sustainable inclusive economies.

Learn more at www.idrc-crdi.ca/en.

S. Bery et al. (eds.), *Navigating Challenges for Sustainable Growth*,
https://doi.org/10.1007/978-981-97-7894-2

Global Development Network (GDN)

The Global Development Network (GDN) is a public international organization that supports high quality, policy-oriented, social science research in Lowand Middle-Income Countries (LMICs), to promote better lives. GDN promotes research on the premise that contextualised and locally driven research leads to more informed policies, increased policy ownership, better informed implementation, and more sustainable and inclusive development choices. GDN also enables research capacity strengthening across countries and disciplines.

Learn more at www.gdn.int.

Appendix 3
Acknowledgements

Overall Leadership

Mr Suman Bery, Vice Chairman, NITI Aayog
Dr Kapil Kapoor, Regional Director South Asia,
IDRC Prof Jean-Louis Arcand, 5 President, GDN

Overall Coordination

Mr Liankhankhup Guite, NITI Aayog
Mr Sabyasachi Upadhyay, NITI Aayog
Ms Rupal Jain, NITI Aayog

Report Launch Event

Ms Urvashi Prasad, NITI Aayog
Ms Geetu Makhija, NITI Aayog

Team Leaders

Ms Anna Roy, NITI Aayog
Mr Vikas Kumar, IDRC
Ms Ramona Angelescu Naqvi, GDN

IT Oversight

Mr Rajesh Sharma, GDN

Communications Outreach

Mr Yugal Kishore Joshi, NITI Aayog
Ms Keerti Tiwari, NITI Aayog
Ms Anisha Bhasin, NITI Aayog
Ms Subhashree Pati, NITI Aayog
Ms Ragini Singh, GDN
Ms Kanika Jha Kingra, IDRC

S. Bery et al. (eds.), *Navigating Challenges for Sustainable Growth*,
https://doi.org/10.1007/978-981-97-7894-2

Finance Oversight

Mr Naushad Khan, GDN

Logistics Management and Support

Mr K.S. Rejimon, NITI Aayog
Mr Dominic Massey, GDN

Notes and Session Summaries

Dr Shashank Shah, NITI Aayog
Mr Balasubramanyam Pattath, GDN
Mr Vikas Kumar, IDRC

Compilation, Editing and Design of Conference Publication

Ms Urvashi Prasad, NITI Aayog
Dr Shashank Shah, NITI Aayog
Mr Vikas Kumar, IDRC
Mr Balasubramanyam Pattath, GDN

Appendix 4
About the Speakers

(In alphabetical order by first name)

Albert Van Jaarsveld
Director General of the International Institute for Applied Systems Analysis (IIASA)
(Austria)

Albert van Jaarsveld was appointed as the 11th Director General of the International Institute for Applied Systems Analysis (IIASA) in 2018. Prior to joining IIASA, he served as Vice-Chancellor and Principal of the University of KwaZulu-Natal in South Africa, and President and CEO of the South African National Research Foundation (NRF). He received his doctorate in zoology (University of Pretoria), pursued post-doctoral studies and research in conservation biology and global security in Australia and the United Kingdom, and underwent leadership training at the University from Harvard. His research has focused on biodiversity, conservation planning, biodiversity and climate change, and ecosystem services. He was appointed Full Professor at the Universities of Pretoria and Stellenbosch and has published over 100 primary research papers, including highly cited works in science and nature. He has served as Co-Chair of MEA Follow-up: Sub-Global Assessments, member of the International Council for Science (ICSU) Nominations Committee, IPBES Focal Point, chaired the Senior Officials Group of G8 Science Ministers on Global Research Infrastructure, Chair of IGFA, Co-Chair of Belmont Forum, Member of ICSU Review Board (2013),

S. Bery et al. (eds.), *Navigating Challenges for Sustainable Growth*,
https://doi.org/10.1007/978-981-97-7894-2

IPBES External Review Board (2018), Member of external review panel of Future Earth (2020) and ISC Commission on Missions for Sustainability (2021-ongoing).

Alicia Garcia-Herrero
Chief Economist, Asia Pacific Natixis
(Spain)

Alicia is the Chief Economist for Asia Pacific at French investment bank Natixis, based in Hong Kong and an independent Board Member of the AGEAS insurance group. Alicia also serves as a non-resident Senior Follow at the East Asian Institute (EAI) of the National University Singapore (NUS) and Adjunct Professor at the Hong Kong University of Science and Technology (HKUST). In addition, she is a Member of the Board of the Center for Asia-Pacific Resilience and Innovation (CAPRI), a member of the Council of Advisors on Economic Affairs to the Spanish Government, a member of the Advisory Board of the Berlinbased Mercator Institute for China Studies (MERICS) and an advisor to the Hong Kong Monetary Authority's research arm (HKIMR). In previous years, Alicia held the following positions: Chief Economist for Emerging Markets at Banco Bilbao Vizcaya Argentaria (BBVA), Member of the Asian Research Program at the Bank of International Settlements (BIS), Head of the International Economy Division of the Bank of Spain, Member of the Counsel to the Executive Board of the European Central Bank, Head of Emerging Economies at the Research Department at Banco Santander, and Economist at the International Monetary Fund. As regards her academic career, Alicia has served as visiting Professor at John Hopkins University (SAIS program), China Europe International Business School (CEIBS) and Carlos III University.

Amitabh Kant
G20 Sherpa of India
(India)

Amitabh Kant is a governance reformer and a public policy change agent for India, having driven key reforms and initiatives during his tenure as the Chief Executive

Officer of the NITI Aayog (2016–2022) and the Secretary of DIPP (2014–2016), Govt. of India. He has been a key driver of flagship national initiatives such as Startup India, Make in India, Incredible India, Kerala: God's Own Country and the Aspirational Districts Program.

NITI Aayog is India's apex policymaking institution, with the Prime Minister as its Chairman. As CEO of NITI Aayog, Kant has driven a vast range of national-level developmental and policy initiatives which catalyzed India's social and economic development and have brought about a paradigm shift in policy-making.

As Secretary, DIPP, he has driven the Start-up India movement which has led to India emerging as the third-best ecosystem for startups globally. His focus has been to facilitate Ease of Doing Business (EoDB) through predictability, consistency of policies and elimination of rules, regulations and procedures. This led to India jumping 79 positions in Ease of Doing Business Indicators. He also initiated competition and ranking amongst Indian states based on their EoDB indicators.

Arunabha Ghosh
CEO, Council on Energy,
Environment and Water
(CEEW)
(India)

Arunabha Ghosh is a highly respected public policy expert, author, columnist, and institution builder. He is the founder-CEO of the Council on Energy, Environment and Water (CEEW) in India, since 2010. Under his leadership, CEEW has emerged as one of Asia's foremost policy research institutions and has been recognised among the world's top 20 climate think tanks. With a vast experience spanning 48 countries, Arunabha has held positions at prestigious institutions such as Princeton, Oxford, UNDP (New York), and WTO (Geneva).

Currently, he serves on the Government of India's G20 Finance Track Advisory Group and provides guidance to the Sherpa Track for India's G20 Presidency in 2022–23. In 2022, he was appointed by the UN Secretary-General to the High-level Expert Group on the Credibility and Accountability of Net-Zero Announcements by Non-State Actors.

Arunabha's contributions extend to advising India's Prime Minister's Office, ministries, state governments, and international organizations on a wide range of subjects. He has been invited by the Government of France as a Personnalité d'Avenir to provide counsel during the COP21 climate negotiations and has played a key role in HFC negotiations under the Montreal Protocol. He has served on the Executive

Committee of the India-U.S. PACEsetter Fund and was a member of the Environment Pollution (Prevention & Control) Authority for the National Capital Region (2018–2020).

Ashima Goyal
Professor, Indira Gandhi Institute of Development Research (IGIDR) (India)

Ashima Goyal, emeritus professor IGIDR, Mumbai, is widely published in international finance and governance, has received national and international awards, edits a Routledge journal, is active in the domestic policy debate, and has served on several boards and policy committees including the Prime Minister's Economic Advisory Council. Currently she is a member of the RBI Monetary Policy Committee and chair of a task force on repurposing the international financial architecture in T20.

Her research has received national and international awards, including two outstanding research awards from GDN in Tokyo (2000) and Rio de Janeiro (2001); was selected as one of the four most powerful women in economics, a thought leader, by Business Today (2008); was the first Professor P.R. Brahmananda Memorial Research Grant Awardee; received the SKOCH Challenger Award for Economic Policy (2017); and select as one of the most powerful women in Indian business in 2021 and 2022 by the editorial team at Business Today.

Debjani Ghosh
President, National Association of Software & Services Companies (NASSCOM) (India)

Debjani Ghosh is the President of the National Association of Software & Services Companies (NASSCOM) since April 2018. A veteran of the technology industry, she is the fifth president of NASSCOM and the first woman at the helm. NASSCOM, is the premier trade body of the industry in India, with over 2800 Indian and multinational member companies. Guided by the vision of the country to adopt and integrate digital technologies, NASSCOM has been actively working towards enabling a digital transformation in the country through technology integration. As President of NASSCOM, Debjani is responsible for establishing new growth areas for the technology industry in India and works with governments and industry stakeholders to establish policies and initiatives that help accelerate the growth of the sector in India and across the world.

Before joining NASSCOM, she was the first woman to lead Intel India and the Manufacturers' Association for Information Technology (MAIT), she was also a member of the executive council at NASSCOM.

In January 2018, she was felicitated by the President of India under the auspices of the 'First Ladies' program, which honors exceptional women pioneers in their respective fields. She has also been listed in the '100 Most Influential Woman in UK-India Relations: Celebrating women' list.

Debjani holds a bachelor's degree in political science from Osmania University in Hyderabad, India, and an MBA in marketing from S.P. Jain Institute of Management and Research in Mumbai, India.

François Bourguignon
GDN, Board Chair; Prof Emeritus, Paris School of Economics; and former Chief Economist, World Bank (France)

François Bourguignon is Chair of the GDN Board. He is also emeritus professor of economics at the Paris School of Economics. He has been the director of the Paris School from 2007 to 2013. Before that he was the chief economist and senior vice-president of the World Bank in Washington. He spent most of his research career as a professor at the Ecole des Hautes Etudes en Sciences Sociales in Paris. His work bears mainly upon inequality and corrective policies in developed and developing countries as well as at the global level. He has authored a large number of academic papers and books. He has received several awards and merits for his works. He is also active in the international development community, lecturing and advising leading international agencies as well as governments

Hanan Morsy
Deputy Executive Secretary and Chief Economist, United Nations Economics Commission for Africa (Ethiopia)

A dynamic Senior Executive with extensive experience of leading top-quality economic research, policy dialogue and development work for international financial institutions, including the International Monetary Fund, the European Bank for Reconstruction and Development, the African Development Bank and the private sector. A global economic and public policy expert who provided quantitative policy analysis to governments around the world on macroeconomic, fiscal and financial issues as well as on private sector development and structural reforms.

Dr. Morsy was the Director of Macroeconomic Policy, Forecasting and Research Department at the African Development Bank (AfDB). Before joining the AfDB, she was the Regional Lead Economist for Southern and Eastern Mediterranean at the European Bank for Reconstruction and Development. Prior to that, she worked at the International Monetary Fund (2003–2012) across different departments including Fiscal Affairs, Middle East and Central Asia, European, and Monetary and Capital Markets, as well as Advisor to Executive Director.

She holds a PhD in Economics from the George Washington University, a Masters in Economics from the University of California and a Bachelor in Economics and Computer Science from the American University in Cairo, Egypt.

Haroon Bhorat
Professor, University of Cape Town (South Africa)

Haroon Bhorat is Professor of Economics and Director of the DPRU. He holds an NRF B2 rating, and with a total citation estimate of over 7800 and an h-index of 47, he is one of the most cited South African economists globally. He currently serves on the Presidential Economic Advisory Council (PEAC), established by President Ramaphosa to generate new ideas for economic growth, job creation and addressing poverty in South Africa. Prof. Bhorat holds the DST/NRF SARChI Chair in Economic Growth, Poverty and Inequality Research. He is a Non-resident Senior Fellow at the Brookings Institution; a Research Fellow at IZA, the Institute for the Study of Labour in Bonn; and is a member of the UCT College of Fellows.

Prof. Bhorat sits on the editorial advisory board of the World Bank Economic Review, and he is a Board Member of the National Research Foundation (NRF) and UNU World Institute for Development Economics Research (UNU-WIDER), previously sitting on the HSRC Board. He has his PhD in Economics through Stellenbosch University, studied at the Massachusetts Institute of Technology, and was a Cornell University research fellow. Prof. Bhorat's commitments at UCT include lecturing Advanced Labour Economics (Hons), and supervision to Honours, Masters and PhD students.

Homi Kharas
Senior Fellow, Center for Sustainable Development, Brookings institution (USA)

Homi Kharas is a senior fellow in the Center for Sustainable Development, housed in the Global Economy and Development program at Brookings. In that capacity, he studies policies and trends influencing developing countries, including aid to poor countries, the emergence of the middle class, and global governance and the G-20. He previously served as interim vice president and director of the Global Economy and Development program.

He served as the lead author and executive secretary supporting the High-Level Panel co-chaired by President Sirleaf, President Yudhoyono and Prime Minister Cameron, advising the U.N. Secretary General on the post-2015 development agenda (2012–2013). The report, "A New Global Partnership: Eradicate Poverty and Transform Economies through Sustainable Development," was presented on May 30, 2013.

Prior to joining Brookings, Dr. Kharas spent 26 years at the World Bank, serving for seven years as Chief Economist for the World Bank's East Asia and Pacific region and Director for Poverty Reduction and Economic Management, Finance and Private Sector Development, responsible for the Bank's advice on structural and economic policies, fiscal issues, debt, trade, governance, and financial markets.

Jayant Sinha
Member of Parliament &
Chair of Parliamentary
Standing Committee on
Finance
(India)

Jayant Sinha is Member, Global Advisory Board, Observer Research Foundation. He is the Chairperson of the Parliamentary Standing Committee for Finance and a Member of Parliament from Hazaribagh, Jharkhand. In the past, he has served as Minister of State for Finance and Civil Aviation. Prior to his career in public service, Mr. Sinha was a venture capitalist. He has degrees from the Harvard Business School, University of Pennsylvania, and Indian Institute of Technology, Delhi.

Jean-Louis Arcand
President, Global
Development Network
(GDN)
(Switzerland)

Professor Jean-Louis Arcand is a Canadian economist and professor of economics at the Graduate Institute of International and Development Studies in Geneva, as well as an affiliate professor at the Université Mohammed VI Polytechnic in Rabat. He is a Founding Fellow of the European Development Research Network (EUDN), a Senior Fellow at the Fondation pour les études et recherches en développement international (FERDI) and has been a Visiting Professor at Renmin University of China in Beijing, Universidade Federal da Bahia and several universities in Africa and the Caribbean. He was assistant and then Associate Professor at the University of Montréal, and Professor at the Centre d'études et de recherches en développement international (CERDI). Jean-Louis holds a PhD in Economics from the Massachusetts Institute of Technology (MIT), an MPhil from Cambridge University and a BA (high honors) from Swarthmore College. He became the president of GDN in January 2023.

Jessica Seddon
Senior Fellow, Yale
Jackson School of Global
Affairs
(USA)

Jessica Seddon's work on environmental governance focuses on how new sources of data can be leveraged to enable new (and more sustainable) ways of interacting with the environment around us. Her career in India and the U.S. spans academic, programme leadership, and strategic advisory roles focused on institutional design for integrating science into policy and social initiatives.

Seddon is a co-founder of The Institutional Architecture Lab (TIAL) and Senior Fellow at Artha Global, a networked policy consulting organization that supports governments in the developing world to design, implement, and institutionalize policy frameworks that promote prosperity, stability, and resilience. She is also an Adjunct Fellow with the Chair in U.S.-India Studies at the Center for Strategic and International Studies (CSIS) and serves on the academic council of the Indian School of Public Policy in New Delhi.

Seddon has most recently focused on governance of various aspects of the atmosphere, from climate change to air quality to climate intervention. She built and led the global air quality program at the World Resources Institute (WRI) and co-chairs the Global Air Quality Forecasting and Information Services initiative of the World Meteorological Organization. She sits on the World Economic Forum Global Futures Council on Clean Air. Prior to joining WRI, Seddon co-founded and led Okapi, an India-based strategy group incubated at Indian Institute of Technology in Madras that focuses on institutional design for social innovation. She has worked with numerous institutions in India, including as visiting fellow at IDFC Institute (Mumbai) and senior fellow at the Center for Technology and Policy, IIT Madras.

Seddon has published book chapters and articles on infrastructure, Indian political economy, information technology and governance, environmental regulation, and other institutional design topics in international academic and policy venues, including Cambridge University Press, the Journal of Development Economics, Stanford Social Innovation Review, Foreign Affairs, Bloomberg Businessweek, and Harvard Business Review. She earned her Ph.D. in political economy from Stanford University Graduate School of Business and her B.A. in government and Latin American studies from Harvard University.

Kapil Kapoor
Regional Director-
Asia, International
Development Research
Centre (IDRC)
(India)

Kapil, an Indian national, is the Regional Director for the Asia Regional Office of the IDRC. Kapil has over 30 years' experience in international development, specializing in Africa and Asia. He was the Director General for Southern Africa, at the African Development Bank, where he was responsible for the Bank's projects and programs across 13 countries in Southern Africa. He has also served as the Director for Strategy and Operational Policies at the Bank, where he led the preparation of the Bank's Long-Term Strategy for Africa, and the Bank's Private Sector Development Strategy.

Kapil has held a series of senior positions with the World Bank Group, including the World Bank's Representative for Uganda and Zambia and the World Bank's Sector Manager for its poverty reduction, economic management and governance programme in Asia.

Kapil holds a PhD degree in Economics and an MBA degree in Finance.

Manjeev Singh Puri
Distinguished Fellow,
Earth Science and
Climate Change,
The Energy and
Resources Institute
(India)

Manjeev Singh Puri joined the Indian Foreign Service in 1982 and has served as Ambassador of India to the European Union, Belgium, Luxembourg, Nepal and as Ambassador/Deputy Permanent Representative of India to the UN. In addition, he has served twice in Germany (in Bonn and Berlin), in Cape Town, Muscat, Bangkok and Caracas.

From 2005 to 2009, he headed the UN-Economic & Social Affairs-Division in the Ministry of External Affairs of India and led the Indian delegation for the first meeting of the Global Forum on Migration and Development in Brussels in July 2007 and the presentation of various reports by India at the Human Rights Council. During 2011–2012, when India served on the Security Council, he was a senior member of its delegation.

Major areas of his experience and professional focus relate to the environment, in particular climate change and sustainable development. He was a lead negotiator for India at the UN on issues relating to the SDGs and at the UN Conference on Sustainable Development held in Rio de Janeiro, Brazil in June 2012. He was a key member of India's delegation at various Climate Change negotiations, including the Conference of Parties of the UNFCCC in Copenhagen in December 2009 and before that at Montreal, Bali, Bonn and Poznan. Furthermore, he was closely involved with India's participation in the G8-G5 Summits from 2005 and he was the point-person on the Indian side at the Major Economies Forum.

Mari Pangestu
Former Managing Director (MD), The World Bank (Indonesia)

Puri has a Master's degree in Management and a BA (Honours) in Economics from St. Stephen's College, Delhi. He is a Distinguished Fellow and on the Advisory Board of TERI.

Mari Pangestu was the World Bank Managing Director of Development Policy and Partnerships. In this role, Ms. Pangestu provided leadership and oversight of the research and data group of the World Bank (DEC), the work programme of the World Bank's Global Practice Groups, and the External and Corporate Relations function. Ms. Pangestu joined the Bank with exceptional policy and management expertise, having served as Indonesia's Minister of Trade from 2004 to 2011 and as Minister of Tourism and Creative Economy from 2011 to 2014.

She has a vast experience of over 30 years in academia, second track processes, international organizations and government working in areas related to international trade, investment and development in multilateral, regional and national settings.

Ms. Pangestu also served as a Senior Fellow at the Columbia School of International and Public Affairs, Professor of International Economics at the University of Indonesia, Adjunct Professor at the Lee Kuan Yew School of Public Policy and Crawford School of Public Policy, Australian National University and a Board Member of the Indonesia Bureau of Economic Research (IBER) and the Centre for Strategic and International Studies (CSIS), Jakarta. She obtained her bachelor's and master's degree in economics from the Australian National University, and her doctorate in economics from the University of California at Davis. She is married and has two children.

Nagesh Kumar
Director and Chief Executive of the Institute for Studies in Industrial Development (ISID) (India)

Prof Nagesh Kumar is the Director and Chief Executive of the Institute for Studies in Industrial Development (ISID), a New Delhi-based public-funded policy think-tank. He is also a Non-Resident Senior Fellow of the Boston University Global Development Policy Centre, Boston, Mass. USA. Prior to joining ISID in May 2021, Prof Kumar served as Director at UN-ESCAP, Bangkok for 12 years. During 2002–09, Prof Kumar served as the Director-General of the Research and Information System for Developing Countries (RIS), a policy think-tank of the Indian Government. Dr Kumar has also served as an Economist at UNU/MERIT, Maastricht, the Netherlands during 1993–98. Prof Kumar has served on the Boards of the Export-Import Bank of India, ICTSD Geneva, and SACEPS Kathmandu, and as a consultant for the World Bank, ADB, ILO, and UNCTAD, among other international organizations. A PhD from the Delhi School of Economics, he received the Exim Bank's first International Trade Research Award and the GDN's Research Medal. He has authored 18 books and over 120 peer-reviewed papers.

Nandan Nilekani
Chairman and Co-founder, Infosys Ltd., Bangalore and Founding Chairman UIDAI (Aadhaar) (India)

Nandan Nilekani is the Co-Founder and Chairman of Infosys Limited. He was the Founding Chairman of the Unique Identification Authority of India (UIDAI) in the rank of a cabinet minister from 2009–2014. Nandan has also co-founded and is the Chairman of EkStep Foundation, a not-for-profit effort to create a learner centric, technology-based platform to improve basic literacy and numeracy for millions of children. In Jan 2023, he was appointed as the co-chair of the "G20 Task Force on Digital Public Infrastructure for Economic Transformation, Financial Inclusion and Development".

Born in Bengaluru, Nilekani received his Bachelor's degree from IIT, Bombay. Fortune Magazine conferred him with "Asia's Businessman of the year 2003". In 2005 he received the prestigious Joseph Schumpeter prize for innovative services in economy, economic sciences and politics. In 2006, he was awarded the Padma Bhushan. He was also named Businessman of the year by Forbes Asia. Time magazine listed him as one of the 100 most influential people in the world in 2006 & 2009. Foreign Policy magazine listed him as one of the Top 100 Global thinkers in 2010. In 2014, He won The Economist Social & Economic Innovation Award for his leadership of India's Unique Identification initiative (Aadhaar). In 2017, he received the Lifetime Achievement Award from E & Y. CNBC-TV 18 conferred him the India Business leader award for outstanding contributor to the Indian Economy and he also received the 22nd Nikkei Asia Prize for Economic & Business Innovation 2017. He has been inducted as International Honorary Member of the American Academy of Arts and Sciences in 2019.

Nandan Nilekani is the author of "Imagining India", co-author of "Rebooting India: Realizing a Billion Aspirations" and "The Art of Bitfulness: Keeping calm in the digital world."

N. K. Singh
Chairman Finance Commission, and President, Institute of Economic Growth (India)

N. K. Singh is a prominent economist, academician, and policymaker with a notable career in India. He currently holds the positions of President of the Institute of Economic Growth and Chairman of the 15th Finance Commission. Prior to this, he chaired the Fiscal Responsibility and Budget Management Review Committee (FRBM). Mr. Singh served as a member of the Rajya Sabha, the Upper House of the Parliament, from 2008 to 2014, contributing to various influential Parliamentary Standing Committees.

Before entering politics and fiscal policy leadership, Mr. Singh had a distinguished tenure in the Indian Administrative Services. He held key roles such as Expenditure Secretary, Revenue Secretary, and Secretary to the Prime Minister of India. He played a significant part in India's economic reforms of 1991, leading negotiations with international organizations like the World Bank and the International Monetary Fund (IMF).

As an accomplished author, Mr. Singh has written several insightful books, including Politics of Change, Not by Reason Alone, The New Bihar: Rekindling Governance and Development, and his autobiography Portraits of Power: Half a Century of Being at Ringside. His latest book, Recalibrate: Changing Paradigms, features essays covering a wide range of subjects in the context of the COVID-19 pandemic. Mr. Singh has also contributed as a columnist in esteemed Indian newspapers such as Hindustan Times, Hindustan, The Indian Express, The Hindu, and Mint. Currently, he serves as a Special Invitee to the CEEW Board.

Otaviano Canuto
Senior Fellow, Policy Center for the New South (Brazil)

Otaviano Canuto is currently a Senior Fellow at Policy Center for the New South and a Non-Resident Senior Fellow at Brookings Institute. Before that, Mr. Canuto worked at multilateral institutions for 15 years. As the Vice President for Poverty Reduction and Economic Management of the World Bank Group, he and his team were responsible for policy advice in areas such as trade, public sector management and governance, public debt management, gender equality, and poverty reduction. As the Vice President for Countries at the Inter-American Development Bank, he was responsible for client management and relationships with member governments. Canuto also occupied positions as an Executive Director of the boards of the International Monetary Fund (IMF) and of the World Bank, where he overviewed operations and policies implemented by both institutions. Mr. Canuto has also been Deputy Minister for international affairs at Brazil's Ministry of Finance, as well as a professor of economics at the University of São Paulo (USP) and the University of Campinas (UNICAMP).

Paul Samson
President, Center for International Governance Innovation (CIGI), (Canada)

Mr. Samson has more than 30 years of experience across a range of global policy issues, working with international partners from around the world. He is currently focused on the transformation of the global economy through digitisation, scenarios for an evolving world order and institutional global governance challenges. During the 24 years with the Government of Canada, Paul's positions included Director General of Strategic Policy at the former Canadian International Development Agency and Assistant Deputy Minister-level roles with Global Affairs Canada and with International Trade and Finance, Finance Canada. At the Privy Council Office, he held several positions during the tenure of three different prime ministers. He also previously served on the Board of Directors for the Centre for International Governance Innovation. Before completing his doctorate (1996) and MA (1991) in International Relations at the Geneva Graduate Institute, Paul earned a BA at the University of British Columbia. He completed postdoctoral studies in global environment assessment at Harvard University.

Peter Drysdale
Emeritus Professor of Economics and Head of the East Asian Bureau of Economic Research, Australian National University (Australia)

He is widely recognised as the leading intellectual architect of APEC. He is the author of a number of books and papers on international trade and economic policy in East Asia and the Pacific, including his prize-winning book, International Economic Pluralism: Economic Policy in East Asia and the Pacific. He is the recipient of the Asia Pacific Prize, the Weary Dunlop Award, the Japanese Order of the Rising Sun with Gold Rays and Neck Ribbon, the Australian Centenary Medal and a member of the Order of Australia.

Poonam Gupta
Director General,
NCAER & Member of
the Economic Advisory
Council to the Prime
Minister (EAC-PM)
(India)

Poonam Gupta is the Director General of National Council of Applied Economic Research (NCAER) and a member of the Economic Advisory Council to the Prime Minister (EAC-PM). Before joining NCAER, she was the Lead Economist, Global Macro and Market Research at the International Finance Corporation; and the Lead Economist for India at the World Bank. Before this, she has been a Reserve Bank of India Chair, professor at the National Institute of Public Finance and Policy, professor at the Indian Council for Research on International Economics Relations (ICRIER), Associate Professor in the Department of Economics at the Delhi School of Economics, and an Economist at the International Monetary Fund. Her research has been published in leading scholarly journals and featured in The Economist, The Financial Times, and The Wall Street Journal. She holds a PhD in International Economics from the University of Maryland, USA and a Masters in Economics from the Delhi School of Economics, University of Delhi.

Ram Madhav
President,
India Foundation (IF)
(India)

Ram Madhav is an Indian politician, social leader, author and thinker. In over a decade of India Foundation's existence, Dr Madhav has been the curator of major annual global and national multilateral initiatives like the Indian Ocean Conference, the Dharma-Dhamma Conference, ASEAN-India Youth Summit and Counter Terrorism Conference involving heads of nations and leaders of governments besides academics, scholars and public-spirited individuals. Most recently, Dr Madhav has been instrumental ideating the Religon-20 Forum (R20) as part of India's presidency of the G20.

Previously, Dr Madhav has served as the National General Secretary of the Bharatiya Janata Party (BJP) during 2014–20 responsible for handling the political affairs of Jammu & Kashmir, Assam and other North-Eastern states of India.

A renowned author and thinker, Dr Madhav has over 300 publications to his credit. He has authored several books in English and Telugu that include "The Hindutva Paradigm—Integral Humanism and the Quest for a Non-Westernw worldview"; "Because India Comes First: Reflections on Nationalism, Identity and Culture" and "Uneasy Neighbors: India and China after 50 years of the war".

Ramesh Chand
Member, NITI Aayog
(India)

Ramesh Chand is currently Member, NITI Aayog, in the rank and status of a Union Minister of State. He has a PhD in agricultural economics from the Indian Agricultural Research Institute (IARI), New Delhi. He is a Fellow of the National Academy of Agricultural Sciences and the Indian Society of Agricultural Economics. He has been involved in policy formulation for the agriculture sector for the past two and a half decades. Prior to joining NITI Aayog, he was Director, National Institute of Agricultural Economics and Policy Research, New Delhi.

Prof. Ramesh Chand has worked in senior academic positions across India, Australia, and Japan. He has also been a consultant with international organisations such as FAO, UNDP, ESCAP, UNCTAD, Commonwealth, and the World Bank. Prof. Chand has chaired important committees on food and agricultural policies set up by various Ministries of the Government of India. He has served as India's nodal officer for agriculture for SAARC for 7 years and represented the country in meetings of G20, UNESCAP.

He has been presented with the Jawaharlal Nehru Award (1984), Rafi Ahmad Kidwai Award (2006) of the Indian Council of Agricultural Research, and the Atal Bihari Vajpayee Award (2018) by the Indian Economic Association.

Robert Lawrence
Albert L. Williams
Professor of International
Trade and Investment
Harvard University
(USA)

Senior Fellow at the Peterson Institute for International Economics, and a Research Associate at the National Bureau of Economic Research. He currently serves as Faculty Chair of The Practice of Trade Policy executive program at Harvard Kennedy School. He served as a member of the President's Council of Economic Advisers from 1998 to 2000. Lawrence has also been a Senior Fellow at the Brookings Institution. He has taught at Yale University, where he received his PhD in economics. His research focuses on trade policy. His publications include Crimes and Punishments? Retaliation under the WTO; Regionalism, Multilateralism and Deeper Integration; Single World, Divided Nations?; and Can America Compete? He is co-author of Has Globalization Gone Far Enough? The Costs of Fragmentation in OECD Markets (with Scott Bradford); A Prism on Globalization; Globophobia: Confronting Fears About Open Trade; A Vision for the World Economy; and Saving Free Trade: A Pragmatic Approach. Lawrence has served on the advisory boards of the Congressional Budget Office, the Overseas Development Council, and the Presidential Commission on United States-Pacific Trade and Investment Policy.

Robert Stavins
A.J. Meyer Professor,
Energy and Economic
Development, Harvard
University
(USA)

Director of the Harvard Environmental Economics Program and Director, Harvard Project on Climate Agreements. He is a University Fellow, Resources for the Future; Research Associate, National Bureau of Economic Research; elected Fellow, Association of Environmental and Resource Economics; Member, Board of Directors, Resources for the Future; and Editor, Journal of Wine Economics. Robert was

Chairman, Environmental Economics Advisory Board, U.S. Environmental Protection Agency. He was a Lead Author, Second and Third Assessment Reports, Intergovernmental Panel on Climate Change, and Coordinating Lead Author, Fifth Assessment Report. His research has examined diverse areas of environmental economics and policy, and appeared in more than a hundred articles in academic journals and popular periodicals, plus a dozen books. He holds a B.A. in philosophy from Northwestern University, an M.S. in agricultural economics from Cornell, and a Ph.D. in economics from Harvard.

Sachin Chaturvedi
Director General
Research and Information System for Developing Countries (RIS)
(India)

He works on issues related to development economics, involving development finance, SDGs and South-South Cooperation, apart from trade, investment and innovation linkages with special focus on WTO.

Professor Chaturvedi has persistently endeavoured to build up institutions and launching of networks, both at national and international levels. He is credited with the launch of Network of Southern Think Tanks (NeST) and Forum for Indian Development Cooperation (FIDC). He has also created "Delhi Process", a major forum for exchange of ideas on South-South and triangular Cooperation.

Professor Sachin Chaturvedi was also the 'Global Justice Fellow' at the MacMillan Center for International Affairs at Yale University (2009–2010) and has served as a Visiting Professor at the Jawaharlal Nehru University (JNU) and was a Developing Country Fellow at the University of Amsterdam (1996), Visiting Fellow at the Institute of Advanced Studies, Shimla (2003), and Visiting Scholar at the German Development Institute (2007).

Currently, Prof. Chaturvedi is also Vice Chairman, Atal Bihari Vajpayee Institute of Good Governance and Policy Analysis; and ex-officio Vice Chairman of Madhya Pradesh State Policy and Planning Commission. He is also Independent Director on the Board of Reserve Bank of India.

Santiago Levy
Non-resident Senior Fellow, Brookings Institution
(Mexico)

Santiago Levy is a nonresident senior fellow with the Global Economy and Development Program at Brookings. He was previously president of the Latin American and Caribbean Economic Association (LACEA). From 2008 to 2018 he was the Vice President for Sectors and Knowledge at the Inter-American Development Bank (IDB). From 1994 to 2000, he served as the deputy minister at the Ministry of Finance and Public Credit of Mexico. He has also held positions across government and academia, including: General Director, Mexican Social Security Institute; President, Federal Competition Commission; Director for deregulation, Ministry of Industry and Trade; Associate Professor of Economics (tenured), Boston University; Economics Professor, Instituto Tecnológico Autónomo de México.

Mr. Levy has received several prestigious awards in Economics and has published six books, 24 articles in academic journals, and 20 book chapters on economic growth and productivity, social policy, informality, education budgetary and tax policy, trade policy reform, rural and regional development, competition policy, labour markets, and policies for poverty alleviation.

B.V.R. Subrahmanyam
CEO, NITI Aayog
(India)

Shri B.V.R. Subrahmanyam joined as Chief Executive Officer of NITI Aayog on 25.02.2023.

An Indian Administrative Service Officer of 1987 batch (Chhattisgarh cadre), Shri Subrahmanyam has held important assignments over the last three decades in Madhya Pradesh, Chhattisgarh, and Jammu & Kashmir, along with a stint at The World Bank. He has been Secretary in the Ministry of Commerce & Industry, Chief Secretary, Jammu & Kashmir, Principal Secretary, Government of Chhattisgarh, and has held positions in the Prime Minister's Office.

Suman Bery
Vice Chairperson,
NITI Aayog
(India)

Shri Suman Bery is currently Vice Chairperson, NITI Aayog, in the rank and status of a Cabinet Minister. An experienced policy economist and research administrator, Mr Bery took over as NITI Aayog Vice Chairperson from 1 May 2022. At the time of his appointment, Mr Bery was a Global Fellow in the Asia Programme of the Woodrow Wilson International Centre for Scholars in Washington D.C. and a non-resident fellow at Bruegel, an economic policy research institution in Brussels. He was also a member of the Board of the Shakti Sustainable Energy Foundation, New Delhi.

From early 2012 till mid-2016, Mr Bery was Royal Dutch Shell's global Chief Economist based in The Hague. In this capacity, he advised the board and management on global economic and political developments. He was also part of the senior leadership of Shell's global scenarios group. During his time at Shell, he led a collaborative project with Indian think tanks (later published) to apply scenario modeling to India's energy sector.

Before his appointment at Shell, Mr Bery served as Director-General (Chief Executive) of the National Council of Applied Economic Research (NCAER) in New Delhi. During his tenure, NCAER greatly extended its global links and was recognised as one of India's leading think tanks by the independent global Think Tank Initiative. In his decade leading NCAER, Mr Bery was at various times member of the Prime Minister's Economic Advisory Council; of India's Statistical Commission; and of the Reserve Bank of India's Technical Advisory Committee on Monetary Policy.

Prior to NCAER, Mr Bery was with the World Bank in Washington D.C. His career at the World Bank spanned research on financial sector development and country policy and strategy, notably in Latin America and the Caribbean. His experience on financial sector reform in Latin America led to an appointment as Special Consultant to the Governor of the Reserve Bank of India between 1992 and 1994.

He has a master's degree in public affairs from Princeton University's School of Public and International Affairs, and an undergraduate degree in philosophy, politics and economics from Magdalen College, University of Oxford.

Surjit Bhalla
Former Executive Director for India, Sri Lanka, Bangladesh and Bhutan, IMF (India)

Surjit S. Bhalla is the former Executive Director for India, Sri Lanka, Bangladesh, and Bhutan at the IMF. He was earlier a member of the Prime Minister's Economic Advisory Council and Chairman of Oxus Research & Investments in New Delhi. Dr Bhalla has worked as a research economist at the Rand Corporation, the Brookings Institution, in the Research and Treasury departments of the World Bank, and as a consultant to Warburg Pincus. He has worked on Wall Street at Deutsche Bank and at Goldman Sachs. He is the author of several academic papers and books: Imagine There's no Country (2002), Devaluing to Prosperity (2012), The New Wealth of Nations (2017), and Citizen Raj (2019). He is a regular contributor to Indian newspapers, magazines, and television on financial markets, economics, politics and cricket.

Dr Bhalla has an MPA and PhD in Economics from Princeton University, and a Bachelor's degree in Electrical Engineering from Purdue University

Tao Zhang
Chief Representative for Asia and the Pacific, Bank of International Settlements (BIS) (China)

Mr. Tao Zhang has been Chief Representative of the BIS Office of Asia and the Pacific since September 2022. As a member of the BIS senior management, he takes lead in its activities in Asia and the Pacific. Mr. Zhang has extensive experiences both in international arena and at the national level in China. He served as Deputy Managing Director of the International Monetary Fund (IMF) in Washington, DC during 2016–2021. In that capacity, he oversaw the Fund's engagement with more than 100 member countries and had a wide portfolio, ranging from financial stability policies, fintech and digital currencies, climate financing and sustainable growth, to engagement with other international organizations. Earlier in his career, he had worked as an economist at the World Bank and the Asian Development Bank. Mr. Zhang also held senior positions in China, including Deputy Governor of the People's Bank of China, and Chairman of the Supervisory Board at the People's Insurance Company (Group) of China Limited. Mr. Zhang has a Ph.D. in International Economics from the University of California, Santa Cruz, USA, and a bachelor's degree from Tsinghua University, China.

Vijay Kumar Saraswat
Member, NITI Aayog

V. K. Saraswat is a highly esteemed scientist and Member of NITI Aayog, renowned for his extensive experience in defense research encompassing both fundamental and applied sciences. Having served as the Secretary of the Defense Research and Development Organization (DRDO), he has made remarkable contributions to the indigenous development of missiles like Prithvi, Dhanush, Prahaar, and Agni-5, as well as the two-tiered Ballistic Missile Defence system. Dr. Saraswat played a pivotal role in establishing infrastructure to support the Nuclear Doctrine, developing cyber security technologies, and enhancing the nation's defense capabilities against ballistic missile threats.

Furthermore, he has been actively involved in the development of alternative energy systems, such as clean coal technologies, concentrated solar power systems, and bioenergy- and hydrogen-based economies. Dr. Saraswat has also made significant contributions to the advancement of technology, including the development of silicon photonics, the configuration of the M-Processor for ICT applications, and the roadmap for the Indian Railway Research Institute.

As a Member of NITI Aayog, Dr. Saraswat has initiated the "Methanol Economy" program, aiming to utilize methanol for transportation, energy generation, and chemical production. Additionally, he has chaired committees on technical textiles and body armor, contributing to the futuristic growth of these sectors in India. Dr. Saraswat's outstanding achievements have been recognised through numerous awards, including the Padma Shri and Padma Bhushan, as well as honorary doctorates from over 25 universities.

Vinod Kumar Paul
Member, NITI Aayog
(India)

Vinod Paul is a Member of the NITI Aayog, where he leads the Health, Nutrition and Education verticals. He has been a key catalyst for a number of flagship schemes of the Government including Ayushman Bharat PMJAY, the world's largest health assurance programme covering over 600 million people.

Dr. Paul has been a part of the core team of the Union Government for Covid-19 pandemic response as the chair of the National Task Force as well as of the National Expert Group on Vaccine Administration for COVID-19 (NEGVAC).

Dr. Paul was conferred with the prestigious Ihsan Dogramaci Family Health Foundation Prize by WHO at the 2018 World Health Assembly for his globally recognised service in the field of family health.

V. Anantha Nageswaran
Chief Economic Adviser,
Ministry of Finance,
Government of India
(India)

Prior to this appointment, Dr. Nageswaran has worked as a writer, author, teacher and consultant. He has taught at several business schools and institutes of management in India and in Singapore and has published extensively.

He was the Dean of the IFMR Graduate School of Business and a distinguished Visiting Professor of Economics at Krea University. He has also been a part-time member of the Economic Advisory Council to the Prime Minister of India from 2019 to 2021. He holds a Post-Graduate Diploma in Management from the Indian Institute of Management, Ahmedabad and a doctoral degree from the University of Massachusetts in Amherst.

MIX
Papier aus verantwortungsvollen Quellen
Paper from responsible sources
FSC® C105338

If you have any concerns about our products,
you can contact us on
ProductSafety@springernature.com

In case Publisher is established outside the EU,
the EU authorized representative is:
Springer Nature Customer Service Center GmbH
Europaplatz 3, 69115 Heidelberg, Germany

Printed by Libri Plureos GmbH
in Hamburg, Germany